The Chemical Industry—Friend to the Environment?

The Chemical Industry—Friend to the Environment?

Edited by
J.A.G. Drake

The Proceedings of a Symposium organised by the Industrial Division of the Royal Society of Chemistry as part of the Autumn Meeting 1991 held on the 24–26 September 1991 at the University of York

Special Publication No. 103

ISBN 0-85186-477-5

A catalogue record for this book is available from the British Library

Published by The Royal Society of Chemistry,
Thomas Graham House, Science Park, Cambridge CB4 4WF

Printed by Redwood Press Ltd, Melksham, Wiltshire

Preface

"The Chemical Industry - Friend to the Environment ?" was a symposium organised by the North East Region committee of the Industrial Division of the Royal Society of Chemistry. It was one of ten symposia held under the auspices of the Royal Society of Chemistry's 1991 Autumn Meeting at the University of York on the 24th, 25th & 26th September. A good attendance at the symposium from the 554 delegates registered for the Meeting as a whole ensured lively discussions after the lectures. This volume contains typescripts from all those lectures.

The general public have mixed views about the Chemical Industry, they appreciate the material comforts it provides for example textiles, ceramics, steel, speciality chemicals, drugs, prosthetics etc. However, many people feel that their comfort is spoiled by the chemical poisoning of the environment through slag heaps, beaches and countryside littered with non-biodegradable unsightly plastic containers, poor air quality through NO_x, CO_2 and chlorofluorocarbon emissions and of course nuclear waste. The occasional spillage of hazardous chemicals through road, rail and sea accidents do nothing to improve the Industry's image.

The majority of these topics were discussed, though no one presumed to know how to remove the problems entirely but many suggestions were put forward as to how this might be achieved. Chemical waste will always be a feature of our lives but determining the fine line between what are acceptable and potentially dangerous thresholds (for example in the case of radioactivity and heavy metals) is a continuing challenge to both the Industry and the public. Recycling has gone beyond the fashionable stage and is now a serious option for plastics. Plastics must be recycled, used as a fuel or be biodegradable. The idea that the polluter must pay is receiving general acceptance but not so readily accepted is the notion that the costs of clean-up may have to be reflected in the price of consumer goods. The co-ordination of emergency services is becoming more easy to achieve through new management techniques which are being used by, for example, BASF.

Many people contributed very generously to the overall success of the Autumn Meeting, especially the speakers. For this symposium special thanks goes to the small organising committee. They were Mr E.I.S. Lee, chairman (BASF plc, Seal Sands), Dr J.P. Candlin (ICI, Wilton), Dr L. Garden (ICI, Wilton), Mr G. Henney, Mr K. Hardy and Dr J.A.G. Drake, secretary, who were also responsible for organising the symposium "Chemicals for the Automotive Industry" in the previous year and whose proceedings have been published in this series by the Royal Society of Chemistry.

Dr J.A.G. Drake, January 1992

Contents

RECYCLING OF PLASTICS

R.D. Rosier

The British Plastics Federation
5 Belgrave Square
London, SW1X 8PD
United Kingdom

1 INTRODUCTION

In 1989 the world wide production of plastics was 1.5 milion tonnes, today it is 75 million tonnes, a fifty fold increase. The plastics industry uses 4% of the crude oil consumption compared with 86% used as fuel in one form or another. There are 3.25 million tonnes of plastics used annually in the United Kingdom of which approximately 1.8 million tonnes finds its way into the waste stream. Of this 1.8 million tonnes, 1.2 million tonnes is from packaging. In this 1.2 million tonnes there are around 600,000 tonnes of polyethylene (PE) film and 250,000 tonnes of bottles which are mainly PE, polyethylene terephthalate (PET) and polyvinyl chloride (PVC). See appendices 1, 2 & 3.

2 SOME DEFINITIONS

As with many industries a jargon has developed within the plastics reclamation industry which can be confusing for those not involved. Three of these terms used are post use, reclaim and recycling. Post use material is that which has performed the task for which it was manufactured. Reclaim is the recovery of post use material from the waste stream, and recycling constitutes converting used materials into another usable material or product.

Post use plastics arise in distinct circumstances which require different approaches for their reclaim, for example:

a) Industrial/commercial. Frequently reasonably large quantities of arisings which are single type plastics that can be clean or only slightly contaminated occur as waste; shrink wrap from supermarkets, garment covers etc.

b) Agriculture. Silage and mulch film, fertiliser sacks etc. The first step here is to arrange for an agent to amass the reclaim from an area in order to make up a worthwhile load for reclaim.

c) Consumer durables and automobiles. There is much discussion, particularly concerning automobiles on this topic; progress will be made, but design for recycling is important. Vehicle batteries are virtually 100% recycled with the lead and the polypropylene (PP) cases being recovered.

d) Domestic waste. Part can and is being recovered by "bring systems", bottle banks etc. with separation of recyclables. Apart from the plastic bottles the plastic occurring is relatively dilute and contaminated which of course constitutes a separation problem.

3 STRUCTURE OF THE PLASTICS RECYCLING INDUSTRY

In the United Kingdom there are some 70 companies which recycle plastics. Many of these companies grew from, and still rely on, the reprocessing of industrial scrap from the plastics manufacturing and conversion processes. It is estimated that 99% of this type of material is reprocessed. It is reprocessed rather than recycled since, as defined earlier, recycled relates to post use material.

The plastics waste arisings from the industrial/commercial area are often clean or nearly clean and can often be segregated into plastic type. If it is relatively clean it does not require a washing process and can be worked on by a recycler who handles the particular plastic.

From agriculture and domestic waste washing is needed as it is for some of the industrial/commercial material; even if it is only to remove the paper label. The recycler's plea is "no labels please". Washing equipment is expensive: a 6,000 tonne per annum plant for washing film costs well over £1 million before buildings and support services are considered.

Certainly in the United Kingdom today there are many companies capable of recycling relatively clean post use material. For contaminated material there are three washing etc. lines for film, all within the British Polythene Industries Group of companies. Reprise, a joint venture company between European Vinyls Corporation (UK) Ltd and PVC Ltd have a sorting and washing plant to separate and wash material from bottles. There are also plants to recover PP from vehicle battery cases and a plant to wash material from heavier items such as bottle crates for recycling into drainage pipe.

There is another route to recycling which is directly from the post use material to the end product. This route has processes which accept a degree of contamination and also accept mixed plastics feedstock; some control on the make up of the mix by polymer type is required. By this route products for wood substitute applications are made, for example benches, road signs, parts for road cones and soil stabilisation systems in temporary car parks etc. In the United Kingdom around 6,000 tonnes per annum is used for these mixed plastics applications.

The most difficult type of waste to process is domestic waste where plastics are dilute and contaminated. However, this is where, not surprisingly, the public are aware of and are concerned about plastics waste. The plastic bottle is the one plastic item most noticeable in the domestic waste stream. The Government White Paper calls for 50% of recyclables in the domestic waste stream to be recycled by the year 2000; this is about 25% of domestic waste. They do not, you will notice, include waste from industrial/commercial sources, but it is the public which votes. Local authorities have only a legal obligation to take away domestic refuse, not process it.

Post use material recycled in the United Kingdom is now in excess of 100,000 tonnes per annum out of the 1.8 million tonnes which are in the waste stream. This 100,000 tonnes comprises of at least:

70,000 tonnes PE film,
15,000 tonnes PP,
2,000 tonnes polystyrene (PS),
1,500 tonnes acrylonitrile butadiene styrene polymer (ABS),
6,000 tonnes mixed plastics.

There are some 40 reclaim schemes operating in the United Kingdom apart from special one to one arrangements made by recyclers. These 40 schemes include kerbside collections, bottle banks, supermarket carrier bags, agricultural bags etc. There is a *potential* for money to be made and for the energy saving balance to be right when reclaiming and recycling from industrial/commercial, agricultural and supermarket carrier bag sources. It is expensive and goes from marginally economic to far from economic as soon as domestic waste is reclaimed. There is a strong argument to back the recovery of energy by incineration for the thin, small item, contaminated, mixed plastic which is in our dustbins. Plastic bottles might look easy to recover but there are 250,000 tonnes of them in the waste stream, if those used in industry are taken out, there are around 160,000 tonnes of bottles to be found in the domestic waste stream. There are three main polymers used in the manufacture of bottles, PE, PET & PVC which need to be sorted, the caps and labels removed, then shredded, washed and dried before regranulation. 350 tonnes of bottles were collected from domestic sources in 1990, it will be 2,000 tonnes in 1991, so there is a long way to go. That 2,000 tonnes is 40 million bottles as there are about 20,000 bottles to the tonne.

When plastics are reclaimed from domestic refuse there is a high cost and one way or another the consumer, who always pays in the end, will have to pay for that reclaim, be it in the form of reclaim for recycling into new plastic items, reclaim as energy recovery or as part of refuse derived fuel. The United Kingdom government is discussing this involved subject with the plastics industry as well as other industries such as glass, aluminium can, etc. The plastics industry is liaising and working closely with the other industries particularly in the high profile packaging area. The German government has set tough recycling targets for packaging materials and the European Community is working on a packaging waste directive. The position throughout Europe on recycling legislation could be said to be chaotic. The United Kingdom goverment is talking to industry, trying to judge the best approach and watching what is happening in Germany and elsewhere in Europe. The 50% target mentioned earlier is in place but there is a need to decide how to fund reclaim from the domestic area, otherwise progress will be zero or minimal on that front.

4 EXAMPLES OF RECLAMATION SCHEMES

At Radcliffe in Manchester the British Plastics Federation (BPF) is a co-sponsor of a trial scheme to extract plastics from untreated domestic refuse; the other sponsors are the Greater Manchester Waste Disposal Authority (GMWDA) and Salford University Business Services (SUBS) who also manage the project. The trial will last for 16 months, to finish in December 1991. Results are encouraging as the target to extract 50% of film at 90% purity has been met. Work is continuing to develop the extraction of plastic bottles. The Department of Trade and Industry have given a grant towards the cost of additional equipment for this project and there appear to be real possibilities for the sale of the technology to the benefit of the participants. At this stage tests suggest that the film extracted will be suitable for mixed plastics recycling but not for recycling to virgin standards.

The second project for the BPF is at Sheffield where as part of the Recycling City scheme the BPF is operating a plastics reclaim trial to assess reclamation from kerbside collection, "bring" systems for bottles and targeted collection from commerical/industrial sources etc. The BPF has rented a building from Sheffield City Council with a manager and assistant manager where sorting is done manually by adults with learning difficulties from Sheffield City Council's Crown Hill establishment and supervised by their staff. The costs of the reclaim operation are met by the BPF and sources of plastics for reclaim are the kerbside collection scheme run by Recycling City Ltd which covers 4,000 households. The BPF is one of the co-sponsors of the scheme and pays for the delivery of plastics for reclaim since it has 45 bottle savers on sites throughout the city as well as other sources and can use its own 7.5 tonne box van. The bottle savers are fitted with re-usable plastic bags for easy manual emptying.

The collected plastics are sorted into type on a picking conveyor line then baled on a 40 tonne pressure horizontal baler. The main type of plastic is PE film and from the bottle stream, PE, PET and PVC. The bales of film weigh approximately 500 kg and the bales of bottles 250 kg; the reduction in volume of the bottles is of the order of 20:1. Sorted material is sold for recycling. Present throughput is 5 tonnes per week of sorted material, with a target of 7 tonnes per week by the end of 1991 rising to 10 tonnes per week early in 1992. As it is a relatively small scheme the economics are poor, however, it does demonstrate the high cost of such collection/reclaim schemes as costs in 1992 are forecast to be in excess of £120,000 with an income from sales of about £35,000.

Another way to recycle plastic is to degrade the plastic to its monomer ready for re-polymerisaton. Work is being undertaken on this aspect in the United States of America on PET bottles where the recovered PET has been recycled back into bottles; a true "closing of the loop". However, the high costs in terms of collection, sorting, processing etc. and the high energy needed make it a challenging project.

5 USES OF RECYCLED PLASTICS

Uses of recycled plastic are increasing while the consumer is pressurising the authorities and industry to recycle plastics. Film products are recycled into carrier bags, domestic waste sacks, pedal bin liners, building film, agricultural film etc. Bottle material is used for 100% recycled bottles in PE and PVC, more frequently as a middle layer in a sandwich between virgin plastic to overcome colour difficulties. Garment hangers are made once more into garment hangers. PET bottles are recycled into duvet fillings, anorak filling, industrial strapping and carpet backing. PP battery cases are recycled back to battery cases. Bottle crates etc. are recycled into drainage pipe and mixed plastics into wood substitute etc. applications described earlier. However, recycled material is only used in applications where it does not come into direct contact with foodstuffs.

6 CONCLUSIONS

Plastics recycling has still a long way to go, but the message is "Plastics can be recycled - plastics are being recycled". Funding of reclaim from domestic sources needs to be resolved; with that and a balance between material recycling, refuse derived fuel and incineration with energy recovery, there is a great future ahead.

APPENDIX 1

UNITED KINGDOM PLASTICS CONSUMPTION BY END-USE

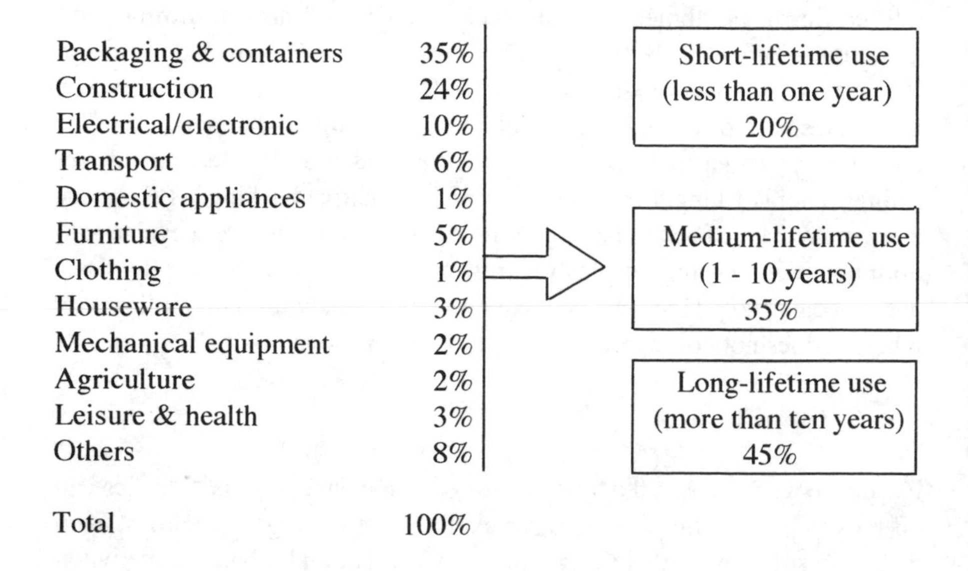

APPENDIX 2

UNITED KINGDOM PLASTICS IN PACKAGING BY MAJOR POLMER (1989)

Low density PE	533,000 tonnes
High density PE	270,000 tonnes
PP	159,000 tonnes
PS	86,000 tonnes
PVC	81,000 tonnes
PET	55,000 tonnes
Expanded PS	22,000 tonnes
Total	1206,000 tonnes

APPENDIX 3

TYPICAL COMPOSITION OF UNITED KINGDOM HOUSEHOLD REFUSE (THE DUSTBIN) 1990*

Material	Weight %
Paper & card	31
Plastic film	4
Dense plastics	4
Textiles	2
Glass	10
Ferrous metal	7
Non-ferrous metal	1
Miscellaneous combustible	6
Miscellaneous non-combustible	2
Putrescible	27
Fines	6

* Warren Spring Laboratory

BIOPOL™ POLYESTER: ICI'S TRULY BIODEGRADABLE POLYMER

J. M. Liddell

ICI Biopolymers and Fine Chemicals
Billingham
Cleveland, TS23 1LB
United Kingdom

1 INTRODUCTION

Synthetic polymers have revolutionised the way we live since their discovery and widescale availability over the past forty years. Through their specific features - versatility, convenience and value they now make up, or are essential components of, most of the articles on which we depend in our everyday lives.

At present, consumer demand has generated a requirement for polymers in excess of 100 million tons per annum. This scale of consumption has, however, been a cause for concern both for the consumer and the plastics industry, through the need for the effective management of post consumer waste and the increased use and dependence on fossil fuels.

Industry has thus been looking at a wide range of methods to reduce the unnecessary use of plastics, to recycle where possible and even to re-use. Alongside and compatible with this, ICI has been researching new and different ways to reduce the environmental impact of plastics.

BIOPOL™ is one result of this research and development effort. It should not be seen as the sole solution for plastics waste management but it does provide and additional option which can be used as appropriate, along with established waste management techniques.

2 MICROBIAL POLYESTERS

BIOPOL™ is ICI's trade name for a range of truly biodegradable polymers which are produced from renewable, agricultural feedstocks. They are thermoplastic polyesters, copolymers of poly(3-hydroxy)butyrate (PHB) and poly(3-hydroxy)valerate (PHV). In the copolymer the 3-hydroxybutyrate (3-HB) and 3-hydroxyvalerate (3-HV) monomer units are incorporated randomly through the polymer chain with hydroxyvalerate levels running from 0 to 30% mole ratio. The basic copolymer structure is shown in Figure 1.

Microbial production of polyesters was first discovered by Lemoigne in 1925[1] and has subsequently been found to be carried out by a wide range of microbial species. In most organisms studied to date, PHB is biosynthesised from acetyl coenzyme A by a pathway involving three enzymes, Figure 2. In vivo, the role of PHB is as a carbon and energy reserve material. The advantage to the organism is that through synthesis of the higher molecular weight polymer, there is an ability to store large quantities of reduced carbon without significantly affecting the osmotic pressure of the cell. As natural materials all the microbial polyesters including the BIOPOL™ copolymers can be degraded in the environment since systems have evolved naturally to handle this type of material.

Since the polyesters are synthesised biologically there is exclusive production of one stereoisomer. In the case of PHB the asymmetric carbon is in the R configuration.

BIOPOL COPOLYMERS

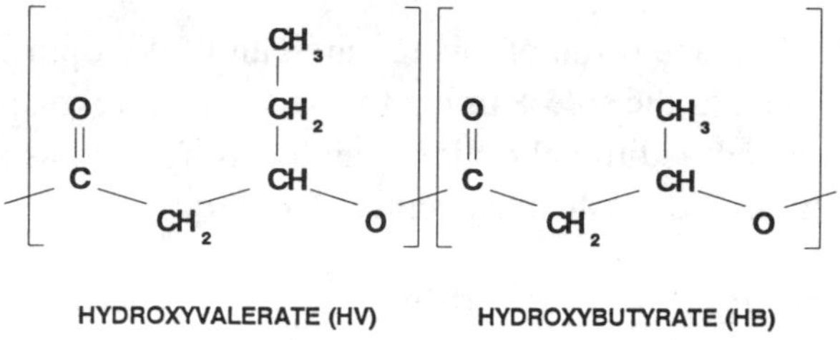

$$\left[-\overset{O}{\overset{\|}{C}}-CH_2-\underset{CH_2CH_3}{CH}-O- \right]\left[-\overset{O}{\overset{\|}{C}}-CH_2-\underset{CH_3}{CH}-O- \right]$$

HYDROXYVALERATE (HV) HYDROXYBUTYRATE (HB)

HV content 0 - 30 mole%

Fig 1

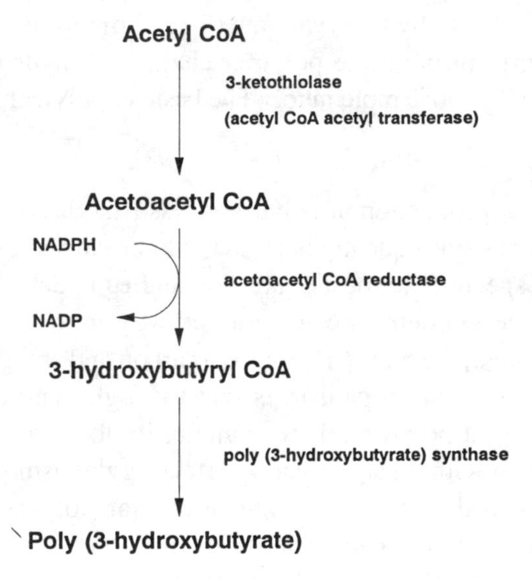

PHB BIOSYNTHETIC PATHWAY

Fig 2

The BIOPOL™ copolymers are examples of a general class of microbial polyesters which contain hydroxyacyl monomer units, the polyhydroxyalkanoates (PHA). PHB is the most abundant of the PHA's in nature but continuing research into the microbial polyester area has lead to the discovery of polyesters other than PHB/V. Investigations involving growing different micro-organisms with different substrates has demonstrated the microbial production of other polyesters. These other polyhydroxyalkanoates fall into two categories.

a) Polyesters of long chain 3-hydroxyalkanoates, Figure 3. (Polyesters having a longer carbon chain have been found from growth of organisms on alkanes or alkanoic acids.[2] These polyesters differ markedly from the PHB/V copolymers being elastomers with a low crystallinity.)

b) Polyesters of 4 & 5-hydroxyalkanoates. (A range of copolymers containing 4HB and 5HB have been formed from growing organisms on organic acid substrates of the same carbon chain length as the desired copolymer.[3] The 3HB - 4HB copolymers (Figure 4) are random copolymers like the 3HB - 3HV copolymers and have been found to have similar material properties.)

Although at present PHA's other than BIOPOL™ are at a very early stage of development, all the PHA's are biodegradable and hence there exists the potential to produce a complete range of polymer material properties from rigid plastics to elastomers which are produced from renewable resources and can be degraded in the environment.

3 BIOPOL™ PRODUCTION TECHNOLOGY

BIOPOL™ PHA is a product of fermentation. A wide range of different organisms have been reported to produce PHB including gram negative, gram positive and cyanobacteria.[4] Hence a wide spread of potential organisms exist as possible production systems for PHB/V.

LONG CHAIN POLY 3-HYDROXYALKANOATES

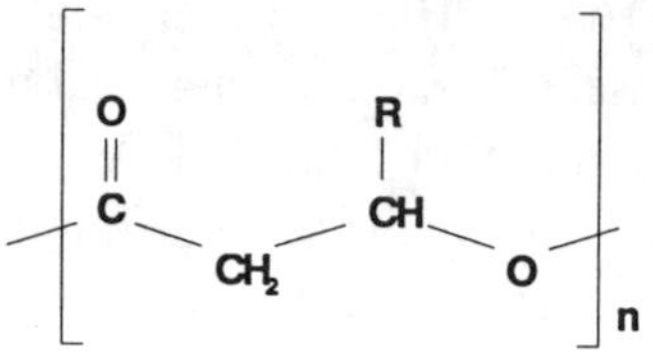

$$\left[-\overset{\displaystyle O}{\overset{\|}{C}}-CH_2-\overset{\displaystyle R}{\overset{|}{CH}}-O- \right]_n$$

R = 6 - 12 chain length

Fig 3

3-HYDROXYBUTYRATE/4-HYDROXYBUTYRATE COPOLYMER

$$\left[-\overset{\displaystyle O}{\overset{\|}{C}}-CH_2-CH_2-CH_2-O- \right]\left[-\underset{\displaystyle O}{\underset{\|}{C}}-CH_2-\underset{\displaystyle CH_3}{\underset{|}{CH}}-O- \right]$$

4-HYDROXYBUTYRATE 3-HYDROXYBUTYRATE

Fig 4

Several factors, however, govern the economics of fermentation processes and require to be considered in development of a fermentation based process. These include:-

a) Product yield.
b) Process complexity.
c) Downstream processing.

These will all have an impact on the capital cost of the fermentation and separation process as well as the operating costs (i.e. raw materials, utilities etc.). The overall production costs will thus be affected substantially by the selection of production organism and substrate.

In the development of the BIOPOL™ process, ICI considered a number of different production organisms and substrates. From experiments or by calculations derived from the metabolic pathways concerned, the yield values for cell biomass and PHB could be determined. Hence an estimate could be made of the cost per unit of PHB produced based on the costs of different substrates.[5] Some of these data are summarised in Table 1.

Table 1 PHB substrate costs for different substrate types.[5]

Substrate	Price (£/t)	Yield (PHB/t substrate)	Substrate Cost (£/t PHB)
Methanol	90	0.18	500
Sucrose	200	0.33	600
Glucose	360	0.33	1100
Hydrogen & carbon dioxide	500	1.0	500
Ethanol	440	0.5	880
Acetic acid	370	0.33	1220

Routes to produce the polymer were assessed and eliminated to arrive at the preferred organism/substrate combination. Among the reasons for eliminating some of the possible routes were:-

a) low polymer molecular mass,
b) unstable polymer production,
c) difficult downstream processing,
d) other co-products,
e) hazardous substrates.

From this type of analysis of potential routes the current fermentation process of the organisms *Alcaligenes eutrophus* grown on glucose was selected. This micro-organism is ubiquitous in the environment and can grow on a wide range of carbon substrates in both aerobic and anaerobic conditions. *Alcaligenes eutrophus* is non-pathogenic and accumulates high levels (up to 80% of the dry cell weight) of high molecular mass polymer.

In the BIOPOL™ fermentation process, *Alcaligenes eutrophus* is grown on a glucose salt medium in a fed batch fermentation process. The fermentation is carried out in two distinct stages.

a) Batch growth of cell biomass.
b) Polymer accumulation.

Exhaustion of a nutrient is used as the growth limiting factor and by the time of growth cessation little PHB synthesis has occurred. At this point glucose is then added to the culture and large accumulation (up to 80% of the dry cell weight) of the polymer occurs.

To produce the co-polymer of PHB and PHV, a similar two stage fermentation is used. However, in this case, in the polymer accumulation phase, a mixture of glucose and propionic acid is added. The hydroxyvalerate content of the polymer is controlled by the ratio of glucose to propionic acid in the feed and hence by controlling the fermentation conditions, ICI can produce a family of biological polymers having a range of properties from the same micro-organism.

The polymer is accumulated as discrete granules within the cell cytoplasm, each cell having a variable number of polymer granules which on average is in the range 8 - 12 (Figure 5).[6] Each granule is thought to be surrounded by a lipid and protein membrane.[7] Recovery of the polymer from within the cell is clearly a vital stage in the overall PHB/V production.

Several methods of extraction of PHB/V have been described including hypochlorite digestion,[8] surfactant pretreatment followed by hypochlorite digestion[9] or extraction by solvents.[10] This last method, involving the use of either chloroform or methylene chloride as extraction solvent was originally investigated by ICI in the early stage of the development of BIOPOL™ but eventually rejected on account of the large volume of solvent required and the high capital investment in recovering the spent solvent from the process.

Instead of the solvent based extraction route, ICI developed an aqueous extraction process for PHB/V which circumvented the problems associated with handling large volumes of solvent.[11] The basic principle of the extraction process is that since PHB makes up the major species in the cell (>70% w/w) the extraction process aims to remove the cell components leaving the purified polymer. To achieve this a series of solubilisation steps are used as shown in Figure 6. Initially the cells are permeabilised, then the cells are treated to achieve solubilisation of cell components. The PHB is separated from the solubilised cell species by centrifugation and washing. The resulting aqueous polymer suspension is then given a final treatment to decolourise the polymer before being dried to a fine white powder. All BIOPOL™ co-polymers are processed by the same basic flowsheet. The resulting polymer powder can then be blended with pigments, fillers or modifiers as required prior to being melt extruded and comminuted to produce polymer chips which are the final product from the process.

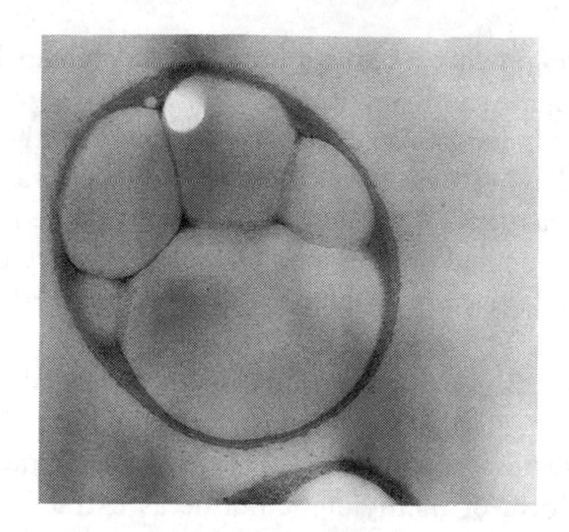

Alcaligenes eutrophus
containing PHB granules
Fig 5

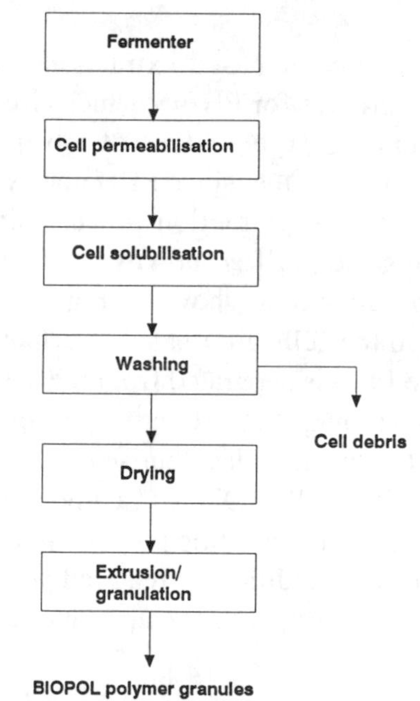

BIOPOL production process
Fig 6

4 PHYSICAL PROPERTIES

The physical properties of the BIOPOL™ co-polymers vary with the 3-hydroxyvalerate content of the polymer. PHB homopolymer is a completely regular polyester and is thus capable of achieving high levels of crystallinity.[12] The homopolymer is a relatively stiff and brittle material. The crystal structure of PHB has been found to consist of polymer chains arranged as compact right handed helices with a two-fold screw axis. A similar structure has been found for pure PHV.[13] The crystal structure parameters are summarised in Table 2.

Table 2 PHB/PHV crystal morphology.

Crystal Morphology	PHB	PHV
Polymer chain configuration	2_1 helix	2_1 helix
Two repeating unit volume (nm^3)	0.11	0.13
Fibre repeat unit (nm)	0.596	0.556

Co-polymers of 3-HB and 3-HV display a melting point minimum at a 3-HV content of approximately 30% mole ratio. For compositions on one side of this minimum 3-HV units crystallise in the PHB lattice while on the other 3-HB units crystallise in the PHV lattice. This phenomena of isodimorphism accounts for many of the physical properties displayed by the co-polymers. In the typical range of 3-hydroxyvalerate levels in the BIOPOL™ co-polymers from 0 - 20%, as the 3-hydroxyvalerate incorporation increases the melting point of the co-polymer decreases from a value similar to polypropylene to a value similar to polyethylene. This allows the co-polymer to be processed at lower temperatures. Flexibility and toughness are improved with increasing 3-HV content. The flexibility increases from a value less than that of polypropylene to a value similar to polyethylene.

In a similar way the tensile strength of the BIOPOL™ co-polymers demonstrates a range of values, which include those of polyethylene and polypropylene, corresponding to the 3-hydroxyvalerate content of the polymer. Typical BIOPOL™ physical and mechanical properties are summarised in Table 3.

Table 3 BIOPOL™ physical properties

HV Copolymer(%)	0	10	20
Melting point (°C)	180	140	130
Tensile strength (MPa)	40	25	20
Flexural modulus (GPa)	3.5	1.2	0.8
Extension to break (%)	8	20	50
Notched Izod impact strength (J/M)	60	110	350

To show how these properties can be exploited in practice, consider the example of a bottle to be made from BIOPOL™. The extrusion blow moulded bottle body needs to be flexible and fully resistant to cracking or splitting when dropped. On the other hand the moulded cap needs to be more rigid to accommodate the thread and to form a good closure. These property requirements are thus met by using a higher 3-HV co-polymer as the principle component of the bottle body formulation whilst a very low 3-HV co-polymer is used to form the cap. The natural physical property range of the BIOPOL™ co-polymers can be further enhanced by the use of normal polymer additives such as modifiers and fillers.

5 BIODEGRADATION

A unique property of BIOPOL™ is that it biodegrades in microbially active environments. Since the polymers are synthesised by micro-organisms as carbon and energy reserve materials, the exercise would be futile if the micro-organisms did not have access to the storage products when necessary. Hence many micro-organisms produce enzymes capable of depolymerising BIOPOL™. There are many examples known of micro-organisms capable of degrading PHB.[14] These micro-organisms secrete an extra cellular depolymerise which degrades and solubilises the polymer in the immediate vicinity of the cell. The soluble degradation products are then absorbed through the cell wall and metabolised.

The rate of degradation is dependent on a number of factors such as environment, temperature, pH, oxygen concentration, surface area, molecular mass and degree of crystallinity. Since many of these factors will change during the course of biodegradation of a particular structure or article it is difficult to give completely definitive rates of biodegradation. The 3-hydroxyvalerate content of the polymer has little effect on the degradation rate. Molecular mass is a more important variable with the rate increasing with decreasing initial molecular mass. Other important factors include the microbial activity of the disposal environment, the level of acidity, temperature, moisture and the presence of other nutrient materials. In aerobic conditions the final products of biodegradation are carbon dioxide, water and a small amount of biological material.

As an example of typical degradation measurements a sample of radio-labelled PHB fabric was buried in a soil in a laboratory test. In 32 weeks a weight loss of 90% was found. Biodegradation was measured by trapping the radio-labelled carbon dioxide gas evolved. It was thus possible to correlate exactly the complete disappearance of the polymer with the gas evolved. The same sample, it should be noted did not biodegrade (and are thus stable) in humid air alone.

In anaerobic conditions the final products of biodegradation are methane and carbon dioxide. Test bars in an experiment were placed in sewage and demonstrated a 90% weight loss in 12 weeks with a near theoretical production of gas. Active anaerobic sewage is a particularly favourable environment for degradation. A thin (100 micron) film immersed in such an environment disappears in approximately one week. Comparative biodegradation data for BIOPOL™ samples in a range of environments are given in Table 4.

Table 4 Biodegradation of PHB in various environments as a 1 mm moulding.

Environment	Time for 100% weight loss (weeks)	Average rate of surface erosion per week (μm)
Anaerobic sewage	6	100
Estuarine sediment	40	10
Aerobic sewage	60	7
Soil	75	5
Sea water	350	1

6 APPLICATIONS

The combination of useful physical properties and the ability to biodegrade indicate a range of potential application areas for BIOPOL™. At one end of the price/volume scale there are a number of interesting medical applications. Since PHB monomer is a normal human metabolic product this taken with the manufacturing method gives BIOPOL™ an apparently high compatibility with mammalian tissue. Combining this with other factors such as the versatility of the BIOPOL™ co-polymers and their slow rate of resorption in the body gives potential applications in the field of slow release and reconstructive surgery.

However, in larger volume markets it is the combination of biodegradability and renewable feedstock which is of most interest. Biodegradability is of most benefit to those products which find their way into sewage systems, soil, compost or landfill sites which are managed for disposal. Applications for this thus lie in the areas of disposable hygiene and agriculture, using either structures made wholly out of BIOPOL™ or degradable composites containing BIOPOL™. Other application areas include packaging, where BIOPOL's™ unique features allow alternative management strategies.

It is now widely recognised that no single solution to waste management is feasible. The order of preferred waste management strategies favoured by the European Economic Community are:-

a) Prevention/reduction.
b) Re-use.
c) Recycle raw materials.
d) Composting.
e) Recycle energy.
f) Safe landfill.

Protection of resources requires both industry and society to work together and providing that re-use and recycling can be achieved in an energy efficient manner then they must be key targets.

It should be noted that BIOPOL™ is believed to be completely recyclable in the conventional sense both by itself and in a mixed plastic waste stream. It is unlikely to have adverse effects on subsequent products both because the proportion of BIOPOL™ waste compared with conventional plastic waste is likely to remain low and because of the polymer's stability during normal storage and use. However, in some circumstances re-use and recycling are unlikely to be viable technically, economically or logistically. In this case composting will provide a solution.

Today more than 50% of the waste stream is degradable and if this were composted it would reduce landfill requirements and allow this organic matter to be recycled to a stable soil-like product. Commercial composting operations exist in both the USA and Central Europe. Plants can have capacities of 10,000 - 20,000 tonnes per annum with the process taking from a few weeks to a few months depending on the technology used and the grade of compost produced. Hence there is also widespread interest in the use of BIOPOL™ in packaging applications where re-use and recycling are difficult or inappropriate but where packaging made with BIOPOL™ could be recycled by composting. Examples include extrusion coated paper/board products, extrusion blow moulded bottles and blown films.

BIOPOL™ polymers can also be incinerated, giving an energy value comparable with traditional plastic materials. Whether the polymers are biodegraded or incinerated, the amount of carbon dioxide released on disposal will be the same as that fixed photosynthetically at the start of the cycle where glucose is used as the fermentation feedstock for BIOPOL™ production.

7 PRODUCTION CAPACITY AND PRICE

As discussed in the previous section there are immediate and developing markets for the BIOPOL™ co-polymers. At present, since BIOPOL™ is still at an early stage of its evolution production capacity is limited. Currently production is in the region of 200 - 300 tonnes per annum. This should rise incrementally, subject to demand, to several thousand tonnes per annum by the mid 1990's.

The cost of production reflects these changes in scale. Production costs have been reduced progressively over the last ten years from around £1000 kg^{-1} in 1980 to £20 kg^{-1} in 1988/89. This trend should continue such that the price for commercial supplies will have fallen to around £3 to £5 kg^{-1} by the mid 1990's. The data in Table 1 indicate that substrate costs can set a lower limit to the production cost of biological polymers.

In the longer term, it may be possible to transfer the necessary genes for PHA synthesis to plants,[14] hence avoiding the fermentation steps and the need to provide fermentation substrate.

8 CONCLUSION

The BIOPOL™ co-polymers are not intended to displace traditional polymers or the re-use, recycle and incineration initiative that are currently underway (in some of which ICI is actively participating through its other product groups). They do, however, provide an additional option to the problems of solid waste and one that is particularly well suited to specific areas of the agriculture and composite packaging materials.

Although still at an early stage in its commercial development BIOPOL™ provides customers and industry alike with a new option, a natural polymer which combines many of the properties and characteristics of traditional thermoplastics, including the ability to recycle with the added features of full biodegradability and manufacture from renewable resources. This advantage has received international recognition in 1991 by the winning of two major environmental awards in Germany and the USA.

9 REFERENCES

1. M. Lemoigne, *Ann. Inst. Past.*, 1925, **39**, 144 - 173.
2. R.G. Lageveen, G.W. Huisman, H. Preusting, P. Ketelaar, G. Eggink, B. Witholt, *Appl. Environ. Microbial*, 1988, **54**, 2924 - 2432.
3. Y. Doi, A. Segawa, M. Kumioka, *Inst. J. Biol. Macromol.*, 1990, **12**, 106 - 111.
4. E.A. Dawes, P.J. Senior, *Adv. Microb. Physiol.*, 1973, **10**, 135 - 266.
5. S. Collins, *Spec. Publ. Soc. General Microbiol.*, 1987, **21**, 161 - 168.
6. D.G.H. Ballard, P.A. Holmes, P.J. Senior, "Recent advances in mechanistic and synthetic aspects of polymerisation", ed M. Fontanille & A. Guyot, Reidel (Kluwer), Lancaster UK, 1987, **215**, 293 - 314.
7. J.M. Merrick, M. Doudoroff, *J. Bacteriol.*, 1964, **88**, 60 - 71.
8. D.H. Williamson, J.F. Wilkinson, *J. Gen. Microbiol.*, 1958, **19**, 198 - 209.
9. J.A. Ramsay, B.A. Ramsay, E. Berger, C. Chavarie, *Biotechnol. Tech.*, 1990, **4**, 221 - 226.
10. P.A. Holmes, L.F. Wright, B. Alderson, P.J. Senior, Imperial Chemical Industries plc, Eur Patent Appln EP15123, 1980.
11. P.A. Holmes, G.B. Lim, Imperial Chemical Industries plc, Eur Patent Appln 8411670, 1980.
12. R. Alper, D.G. Lundgren, R.H. Marchessault, W.A. Cote, *Biopolymers*, 1963, **1**, 545 - 556.
13. T.L. Bluhm, G.K. Hamer, P.R. Sundarajan, *Polym. Prep.*, 1988, **29**, 603 - 604.
14. P.A. Holmes in "Developments in crystalline polymers - 2", D.C. Bassett ed, Elsevier Applied Science, London, 1988.
15. R. Pool, *Science*, 1989, **245**, 1187 - 1189.

BIODEGRADATION OF DETERGENTS

J.S. Clunie

Procter & Gamble Limited
Forest Hall
Newcastle upon Tyne
Northumbria, NE12 9TS
United Kingdom

1 INTRODUCTION

Laundry detergents and Household Cleaning Products are high tonnage consumer products which, after use, end up in the domestic waste water system. It goes without saying that such products must *not* lead to problems either in sewage works, or in the rivers or estuaries, which receive the treated effluent from these works.

The most important constituent of detergents and cleaners are the surfactants - the principal cleaning ingredient in a washing formulation. Surfactants account for up to 20% by weight of a granular detergent and up to 40% by weight of a liquid detergent.

Surfactants, or surface-active agents, are in general organic compounds which:

a) lower the surface tension of the A/W interface, and so promote foaming,

b) lower the interfacial tension of the O/W interface, and so promote emulsification,

c) adsorb strongly at S/L interfaces, and so alter contact-angles, wettability, and thus help stabilise dispersions.

In terms of structure, surfactants are derivatives of hydrocarbons, made by substituting a polar group X into a hydrocarbon chain R to confer water solubility and hence amphiphilic properties. Classification of surfactants is usually based on polar group type, Figure 1.

Commercially important anionic surfactants are soaps and linear alkylbenzene sulphonate (LAS), Figure 2. A recent estimate, by the consulting company A.D. Little, has put LAS consumption in North America, Western Europe and Japan at about 1 m tonnes per annum. Compared with anionics, non-ionic surfactants show greater structural diversity. Important types are the linear alcohol ethoxylates (LAE) containing from 5 to 15 ethoxylate units per mole of alcohol, Figure 3.

The majority of the anionic and non-ionic surfactants, used by consumers of laundry and cleaning products, are disposed of "down the drain" into the sewage system. In the United Kingdom most sewage goes to over 8,500 municipal waste water treatment plants located across the country. These plants, in turn, discharge millions of gallons of treated effluent per day to receiving waters such as rivers and estuaries. In adddition, there is a need to consider direct discharge via sea outlets and also the small proportion of home treatment systems (septic tanks) which discharge their treated effluent directly to subsurface soil systems.

Because of their broad-scale release into the environment, and their high volume usage, it is important to know what happens to surfactants in the environment. This is where biodegradation processes play a major role

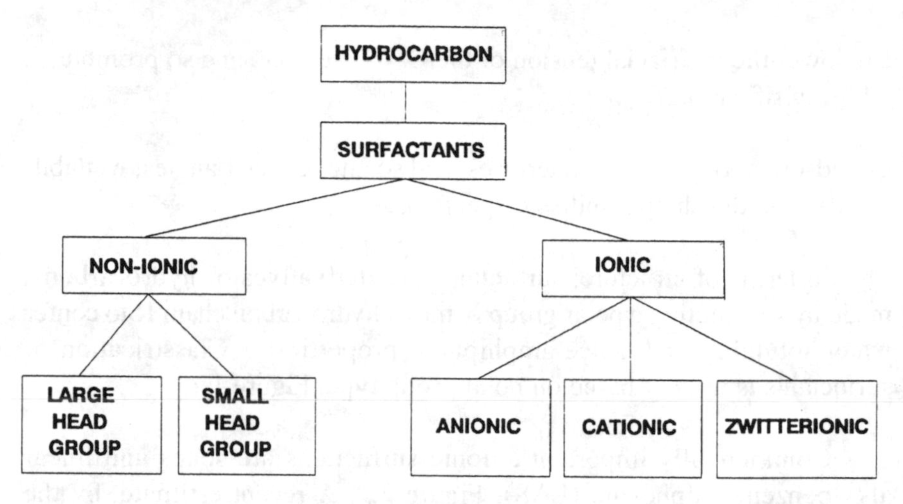

Figure 1

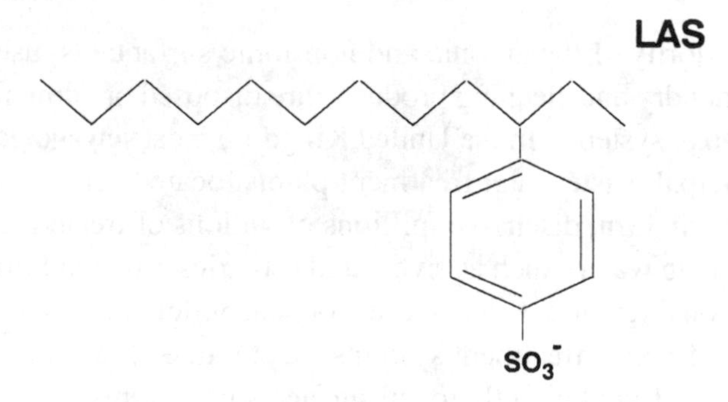

Figure 2

LINEAR ALCOHOL ETHOXYLATE

LAE

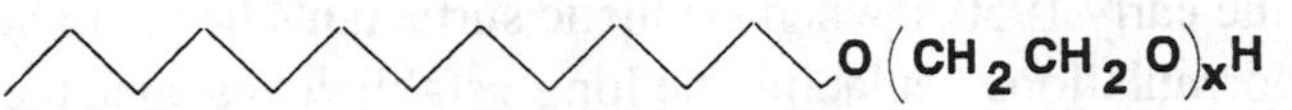

$O(CH_2CH_2O)_xH$

Figure 3

BRANCHED-CHAIN ALKYL BENZENE SULPHONATE

ABS

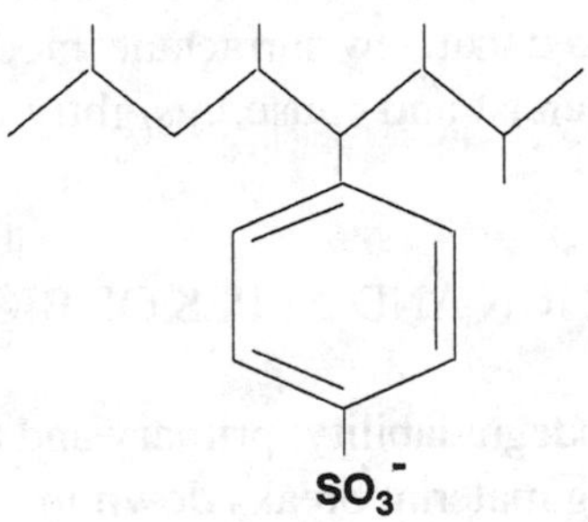

SO_3^-

Figure 4

in preventing the build up or accumulation of surfactants in the environment. Also, biodegradability, especially of surfactants, has recently come more to the fore in the public's imagination, because of a heightened awareness of all things green.

In the early 1950's, when synthetic surfactants began to be used in detergent formulations, replacing the long-established soaps, the alkyl chain was derived from propylene tetramer and was branched; the material was called alkyl benzene sulphonate (ABS), Figure 4. Experience showed the ABS did not biodegrade quickly in sewage treatment plants or in receiving waterways. The starting material retained its foaming properties and generated foam wherever there were turbulent water conditions. Co-operative work by the detergent industry, and its suppliers, led to the development of biodegradable alternatives, and to the voluntary replacement of "hard" ABS by "soft" LAS in 1963. Following this experience with ABS, suppliers have ensured that new surfactants biodegrade to a sufficient degree that they do not persist and cause unsightly foaming problems in natural waterways.

2 BIODEGRADATION AND TYPES OF BIODEGRADATION

There are two types of biodegradability; primary and ultimate. In primary biodegradation the starting material breaks down into smaller components which are unable to react chemically with the analytical reagent used to identify the presence of the parent compound. This co-incides, in surfactants, with the loss of undesirable properties such as foaming.

In "ultimate" biodegradation, the parent compound breaks down completely to the simplest possible materials. A material containing carbon, hydrogen, nitrogen, sulphur and oxygen would biodegrade to carbon dioxide, water and other inorganic compounds like nitrates and sulphates.

The United Kingdom Detergent (Composition) Regulations 1978 and 1984 which implemented the European Directives of 1973 and 1982 require that anionic and non-ionic surfactants, respectively, exceed 80% primary biodegradability in a specified time under specified conditions. The detergent industry had already given a voluntary undertaking to government in 1963 that it would only use surfactants which had greater than 90% primary biodegradability; legislation merely codified the then existing situation.

The most important biodegradability tests specified by European Community detergent legislation are the OECD screening test and the OECD confirmatory test. The first is a simple shake flask test where the surfactant is added as the sole carbon source to a mineral solution. This culture is then inoculated with bacteria.... usually simple addition of some sewage treatment effluent. The fate of the surfactant is then followed by a semi-specific analysis. Anionic surfactants are analysed as methylene blue active substance (MBAS) - involving ion-pair complexation of the anionic surfactant with a cationic dye, chloroform extraction and photometric determination of the coloured complex. Non-ionic surfactants are analysed as a bismuth active substance (BiAS) - involving sublation, ion exchange chromatography, precipitation, re-dissolution and potentiometric titration.

If 80% MBAS or BiAS removal is reached within the official test duration of 19 days, then that surfactant may be used in detergent and cleaning formulations. An example of such a biodegradability determination is shown schematically Figure 5 where two hypothetical materials are compared with a soft LAS standard which passes, and with a hard ABS standard which fails.

The OECD screening test is so stringent that false negatives are to be expected. Those that pass the 80% limit are allowed, those that do not pass are not forbidden but must proceed to the decisive OECD confirmatory test.... which is a simple model of a communal sewage plant. Figure 6 illustrates the United Kingdom porous pot version of the activated sludge simulation test. Here the surfactant to be tested is added to a synthetic sewage. To be considered sufficiently biodegradable, according to the law,

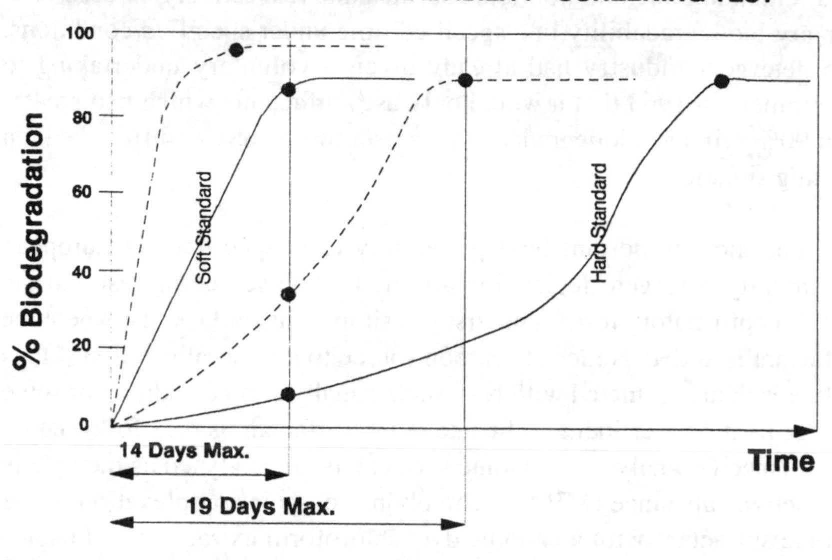

Figure 5

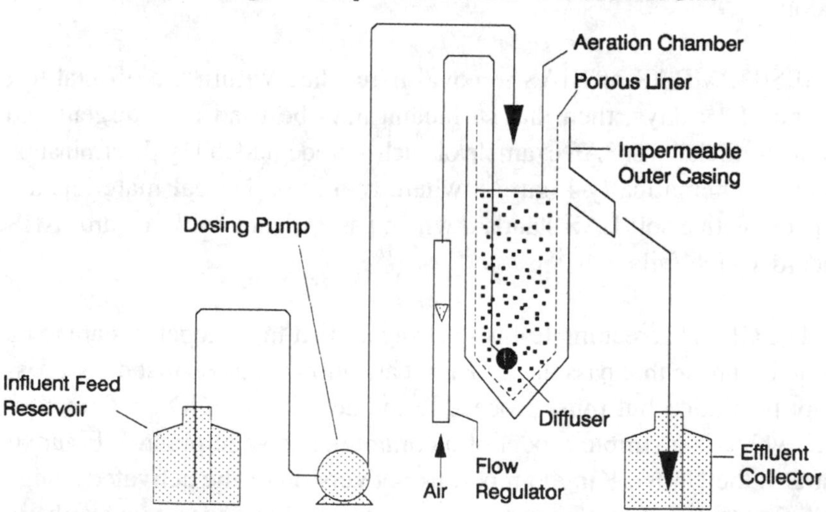

Figure 6

80% removal of the surfactant must be achieved. Surfactants which do not reach this 80% limit are banned from general usage.

As can be appreciated, the Soap & Detergent industry has invested considerable resources to investigating surfactant biodegradability, with much of this effort focused on screening tests. As well as covering the legal aspects, such tests provide useful *comparative* information on the *relative* biodegradability of surfactants, but they suffer from two major limitations:

a) The experimental conditions are conservative; they do not simulate conditions found in the natural environment. This limits the environmental relevance tests and makes it difficult to extrapolate to the "real world".

b) Most screening tests aim to model either sewage sludges or sewage-impacted waters. For the most part, they ignore environmental compartments "beyond the end of the sewage pipe".

Because other non-sewage compartments, such as surface water, soil, sediment, ground water and so forth, all contribute to the overall biodegradation process, Procter & Gamble's Corporate Environmental Safety Department has carried out research to measure surfactant biodegradability in a range of environmental compartments.

3 PROCTER & GAMBLE'S TEST PHILOSOPHY

Biodegradation can be broadly defined as any biological process which converts an organic chemical into organic and/or inorganic end-products which are chemically distinct from the parent material. A more practical working definition of biodegradation would be the metabolism of organic chemicals as a source of carbon and energy by heterotrophic micro-organisms (mainly bacteria) to form microbial biomass and organic and/or inorganic end-products. This latter type of biodegradation, termed ultimate biodegradation (or mineralisation), is highly significant from an environmental viewpoint. It results in a reduction of mass as well as a loss of chemical identity... and plays a major role in preventing the build-up or accumulation of surfactants in the environment.

Although the extent of ultimate biodegradation is important, it is not the most important factor controlling environmental exposure levels. The key factor in controlling the concentration of an organic chemical in the environment is the *rate* of biodegradation over an environmentally relevant time-frame. To obtain meaningful kinetic information, it is important to take measurments on real samples at realistic environmental concentrations. This often requires the use of radiolabelled (mainly carbon-14) materials, or, alternatively the use of sensitive analytical techniques (e.g. HPLC and FAB-MS), which can detect *and* quantify both parent material and biodegradation intermediates.

To determine complete biodegradation, biodegradation assays should also measure *rates* of mineralisation of those structural moieties of the surfactant that are the *least* susceptible to microbial attack. In the case of LAS, the benzene ring is the last structural moiety to be metabolised. Biodegradation assays on LAS therefore follow the disappearance of UV absorbance, or the production of carbon-14 labelled carbon dioxide from carbon-14 ring labelled material, Figure 7. On the other hand, degradation measurements on alkyltrimethylammonium chloride (C_{12}TMAC and C_{18}TMAC), Figure 8, follow mineralisation of the carbon atoms adjacent to the quaternary nitrogen, using either a nitrogen-specific analytical technique, or methyl/C_1-alkyl labelled material. With alcohol ethoxylates the glycol moiety is radio-tagged, Figure 9.

In many respects, however, the focus on mineralisation assays for key structural moieties is doubly restrictive in that it:

a) follows rates of production of end-products, like carbon dioxide, which are a function of the slowest, rate-determining step in the biodegradation pathway and

b) concentrates on the structural moieties least susceptible to microbial attack and therefore likely to be degraded at the slowest rate.

Linear Alkylbenzene Sulphonate (LAS)

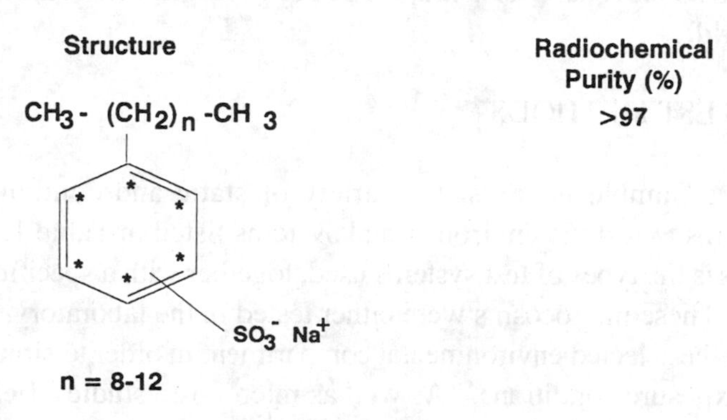

Figure 7

Alkyl Tri-Methyl Ammonium Chloride

(C_{12} TMAC, C_{18} TMAC)

Structure	Radiochemical Purity (%)
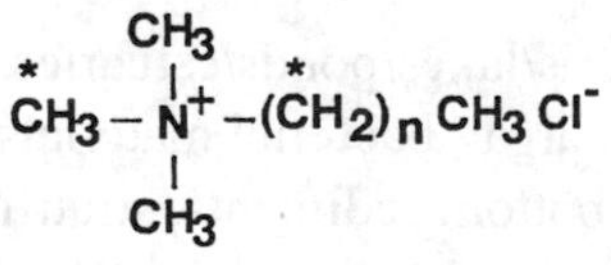 $CH_3-\overset{*}{N}^{+}(CH_3)_2-\overset{*}{(CH_2)}_n\,CH_3\,Cl^-$	>98

n = 11-17

Figure 8

Linear Alcohol Ethoxylate

(C_{12} E_9, C_{15} E_7)

Structure	Radiochemical Purity (%)
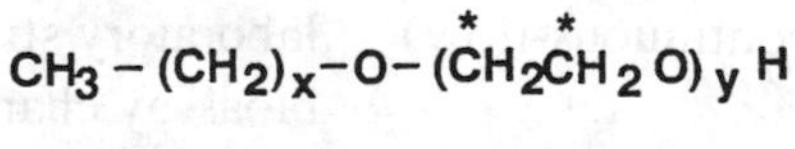 $CH_3-(CH_2)_x-O-(\overset{*}{C}H_2\overset{*}{C}H_2O)_y\,H$	>98

x = 11-14
y = 7-9

Figure 9

On the other hand the advantage of this approach is that it *is* conservative and so prevents over-extrapolation of biodegradation rate data to the "real world".

4 TEST METHODS

Procter & Gamble have used a variety of static and continuous flow microcosms to test the environmental systems listed in Table 1. Table 2 summarises the types of test systems used, together with a specific example of each. These microcosms were either tested in the laboratory, or placed *in-situ* in the selected environmental compartment in order to simulate "real world" exposure conditions. As well as microcosm studies, field studies have been undertaken to assess the fate of surfactants under "worst case" exposure conditions... using natural systems where several environmental fate and biodegradation processes can be integrated.

Table 1 Systems tested in various compartments using micro-organisms

Compartment	System Tested
aquatic	rivers/streams/lakes/ponds/estuaries
epilithic	periphyton, algal, bacterial materials (ponds)
benthic	suspended bottom sediments, stratified sediment cores
terrestrial	agricultural soils, sludge-amended soils
subsurface	ground water, well water, subsurface soil

Table 2 Types of test systems used

Test System	Examples
laboratory microcosm (static)	batch reactor
laboratory microcosm (continuous-flow)	laboratory stream
in-situ microcosm	bioassay chambers
field studies (ecosystem)	lake

The move from laboratory, to *in-situ*, to field systems obviously results in:

a) a higher level of eco-system complexity,
b) more realistic exposure conditions and
c) increased environmental relevance.

The major disadvantages of this progression are:

a) loss of experimental control and flexibility,
b) decreased ability to establish precise cause and effect, or dose/response relationships and
c) a significant increase in the cost and effort to get the desired results.

Biodegradation evaluations involve determining the rate of carbon-14 labelled carbon dioxide production of key structural moieties of individual surfactants. At realistic environmental concentrations in the $\mu g\ l^{-1}$ range for aqueous and $\mu g\ g^{-1}$ range for solid samples, the measured biodegradation rates follow first-order kinetics. In other words, the results can be expressed as half-life values, Figure 10. Thus one half-life value represents the time for 50% biodegradation to occur; put another way, it is the time required to yield half the maximal amount of carbon doxide capable of being produced over the given time period.

5 TEST RESULTS FOR KEY SURFACTANTS

The simplest, most cost-effective way to determine the biodegradability of surfactants in various environmental compartments is from static laboratory microcosms. Several types of static microcosm have been used to test the biodegradability of carbon-14 labelled surfactants. These include closed and flow-through batch reactors, semi-continuous batch reactors, soil columns and sediment microcosms.

One such system, a batch reactor, is shown in Figure 11. The environmental samples used in this test method were collected from several

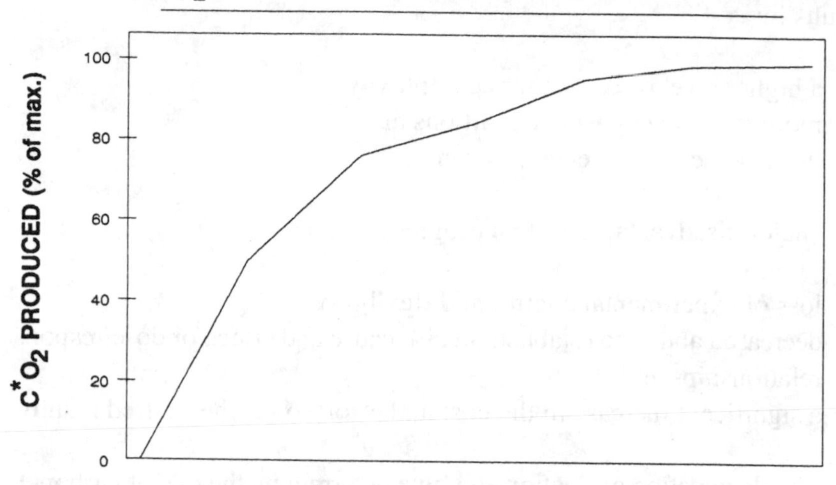

Figure 10

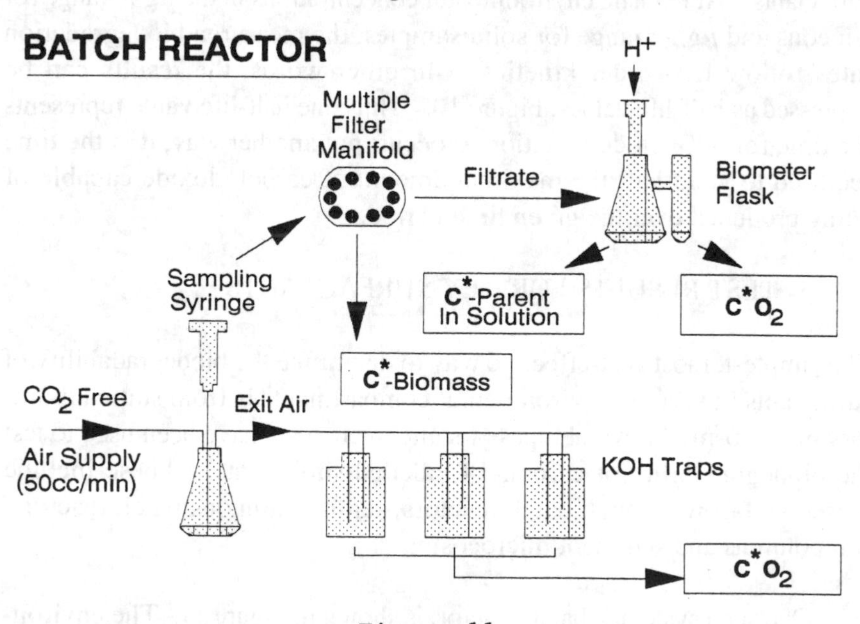

Figure 11

different compartments including river and estuary water, surface and subsurface soils, freshwater, groundwater and estuarine sediments and epilithic environments. All samples were amended with LAS at concentrations approximating realistic environmental levels, namely:

a) 5 - 100 $\mu g\ l^{-1}$ in water and epilithon samples,
b) 10 - 250 $\mu g\ g^{-1}$ in soils and sediments.

As shown in Table 3, LAS degradation was rapid in all environmental compartments tested. Degradation rates were the most rapid in freshwater sediments (half-life less than one day) and slowest in soil sediment (half-life roughly 10 days). Even in soil compartments, degradation half-lives for LAS were comparable with those for naturally occurring compounds as shown in Table 4, where biodegradation half-lives for LAS in a range of soil types are comparable with those for stearic acid (a saturated C_{18} fatty acid found in soaps and cooking oils). Comparable degradation results have likewise been obtained for the cationic C_{18}TMAC and the non-ionic $C_{12}E_9$ surfactants. In general it was found that all three surfactant types were rapidly degraded within a few days in batch reactors using a variety of different environmental samples.

Table 3 LAS degradation in different compartments at different sites

Compartment	Sampling Site	Half-life (days)
river water	Rapid Creek, SD	0.9
estuarine water	Newport River Estuary, NC	7.0
surface soils	Harleysville, PA	10.0
subsurface soils	Summit Lake, WI	5.0
fresh water sediments	Rapid Creek, SD	0.6
ground water sediments	Saulte Ste Marie, ONT	1.5
estuarine sediments	Newport River Estuary, NC	6.0
epilithon	Little Miami River, OH	1.5

Table 4 Degradation of LAS in different soil compartments compared with naturally occurring compounds

Compound	Half-life (days)				
	Silt Loam (average of three)	Loam	Sandy	Organic	Compost
LAS	1.8	3.3	2.6	1.7	1.5
$C_{12}E_9$	2.2	2.7	3.0	1.3	1.7
C_{18}TMAC	5.7	-	5.3	5.8	1.4
stearic acid	2.9	4.1	3.5	-	0.7

Two key factors limit the use of static microcosms:

a) the difficulty in controlling chemical exposure conditions and
b) the difficulty in testing more than one environmental compartment at a time.

In contrast the use of continuous flow or dynamic microcosms largely avoids these difficulties by:

a) allowing exposure conditions to be controlled for extended times and
b) facilitating the dosing and testing of multiple environmental compartments.

Several different kinds of continuous flow microcosms have been used to test surfactant biodegradability, these include:

a) bench-scale wastewater treatment systems,
b) sediment and subsurface chemostats,
c) epilithic and biofilm reactors and
d) model lakes and streams.

Figure 12 illustrates the design basis of Procter & Gamble's model laboratory stream; this is a serpentine 22 m fibre glass channel about 32 cm wide,

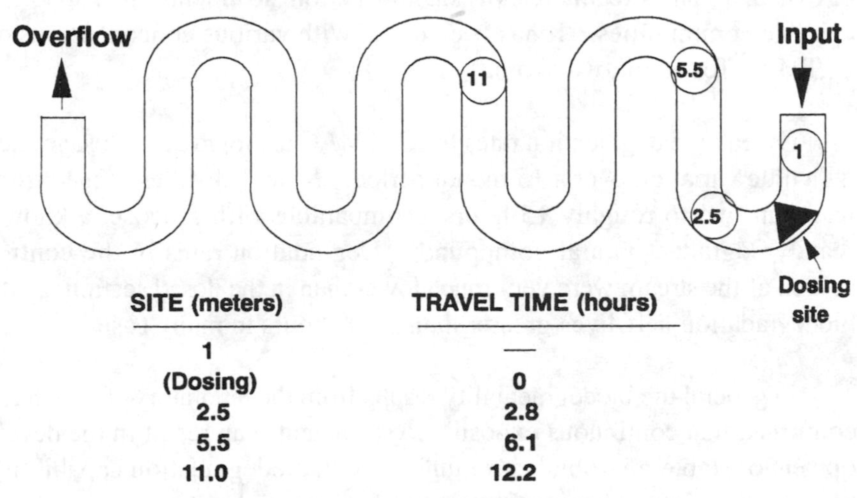

Figure 12

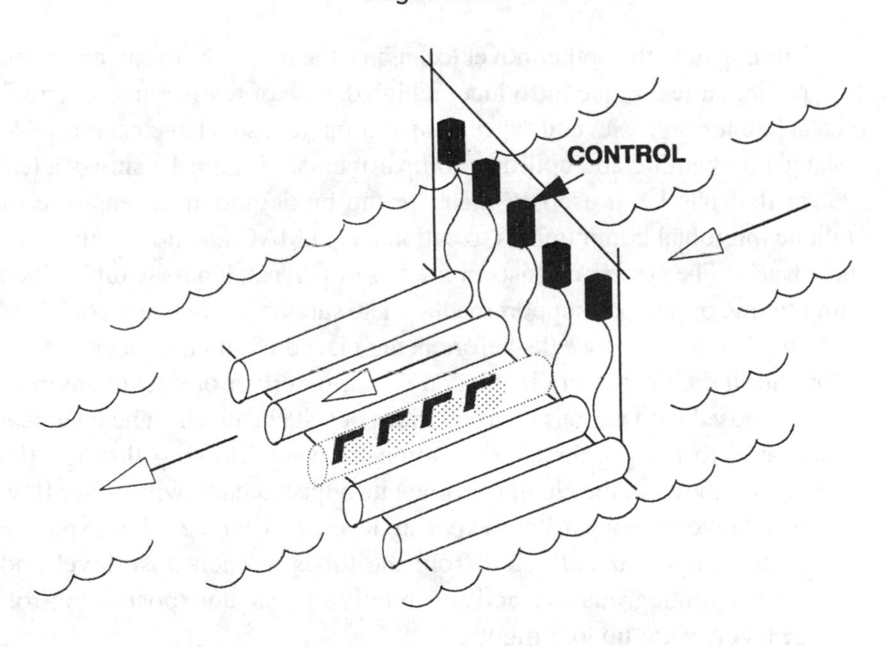

Figure 13

with a water depth of 23 cm and a total volume of 1,500 dm^3. It can be fed with a mixture of river water and well water and operates on a photo-period of 16 hours. The stream contains shallow bottom sediments and supported epilithic communities. It has been dosed with various concentrations of C_{12}TMAC for about two years.

Overall the degradation rates for C_{12}TMAC are rapid at all dosage sites with little variation over a 15 month period. Mean half-lives ranged from less than two to roughly 15 hours - comparable with glucose, a known readily degraded natural compound. Degradation rates in the control section of the stream were very much lower than in the dosed section... with biodegradation half-lives greater than 5,000 hours in many cases.

In general the biodegradability results from the laboratory stream have confirmed that continuous exposure to surfactants can result in the development of stable microbial communities with biodegradation capabilities orders of magnitude higher than those of unexposed communities.

Turning now to another novel technique, the *in-situ* bioassay chamber. This particular technique introduces a high degree of realism into environmental fate testing. Several types of *in-situ* microcosms have been used to test aquatic, benthic and epilithic compartments. Figure 13 shows a test system that has been used to measure the biodegradation response of epilithic microbial communities to cationic C_{12}TMAC and non-ionic $C_{15}E_7$ surfactants. The system consists of a series of perspex bioassay tubes filled with ceramic or perspex support media. The support media were colonised with epilithon for 2 - 4 weeks before being placed within the tubes. After colonisation the tubes were incubated *in-situ* in a river or stream environment and dosed with various concentrations of test chemical. The chemical was diluted to the proper concentration by water flowing through the submerged tubes and the chemical flow rate adjusted daily with water flow rates to achieve nearly constant exposure levels. After a 21 day exposure period, epilithon was collected from the tubes at each dose level and assayed for biodegradation activity relative to an unexposed control. Dosage levels were up to 5 mg l^{-1}.

Across all dosage levels mean half-lives for C_{12}TMAC and $C_{15}E_7$ degradation were rapid, averaging about 2 to 13 hours respectively. These values were among the most rapid measured in any environment compartment and underscore the high degradation activities of epilithic microbial communities for these materials.

In terms of environmental relevance, field studies provide perhaps the most realistic way to characterise the fate of surfactants in natural environments. Given the complexity of most field studies, and the amount of time, effort and money required, field studies can only be used on a limited number of materials. Results from such studies, however, are key to determining what is actually happening in the "real world".

Figure 14 is a schematic diagram of a field site located near Summit Lake, Wisconsin. Wastewater from a local laundromat discharges into a natural depression and this has resulted in the formation of a permanent pond and wetland ecosystem. This wetland system contains *all* of the major environmental compartments discussed earlier (aquatic, epilithic, benthic, soil and subsurface compartments). It has also been exposed to high concentrations of surfactant chemicals, in particular LAS, for more than 25 years. It therefore represents an ideal natural laboratory for studying the biodegradability and fate of surfactants under "worst case" exposure conditions.

Using a mobile laboratory housed on-site to collect and analyse samples, all of the major environmental compartments have been characterised with respect to physical/chemical parameters, surfactant concentrations, microbial ecology and the biodegradative capabilities of indigenous micro-organisms.

The subsurface compartment at Summit Lake is of particular interest because of the potential impact of the laundromat operation on the quality of ground water located beneath the pond. The soil substratum beneath the pond is a porous sand and is in direct communication with the underlying aquifer. Yet significant concentrations of LAS have *not* been detected in

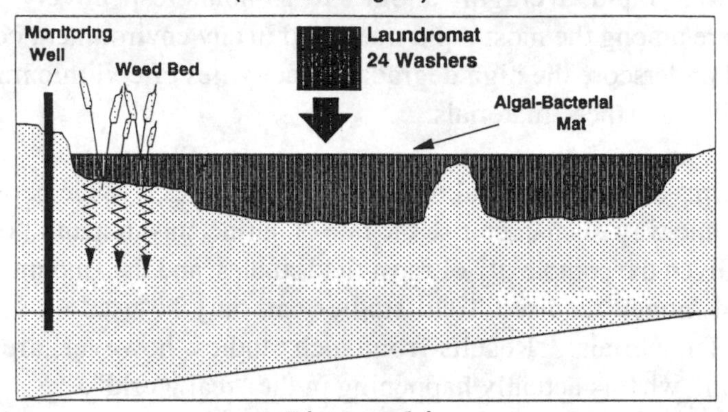

Figure 14

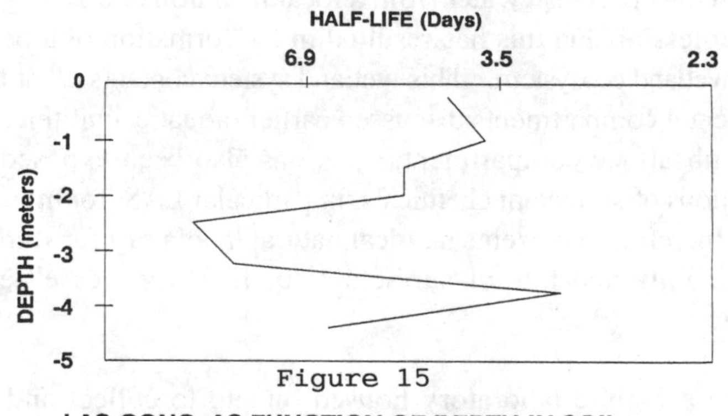

Figure 15

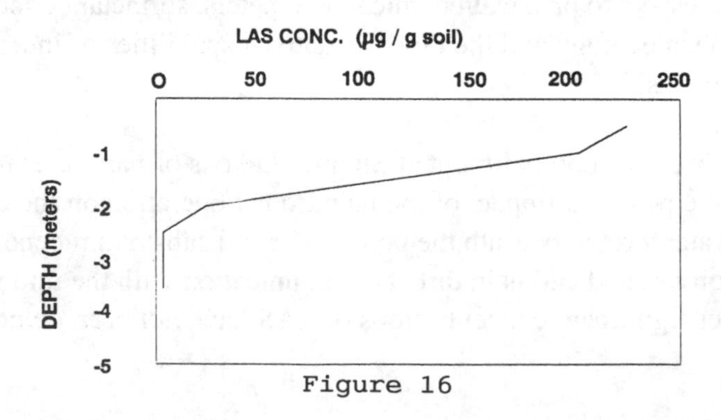

Figure 16

any of several monitoring wells drilled in the area, even though approximately 5,000 kg of LAS have been released to the 25 x 75 m pond since the laundromat operation started. Given the large input of LAS, biodegradation would have to be an effective removal mechanism to prevent LAS accumulation occurring. This hypothesis is confirmed in Figure 15, which charts the biodegradation half-lives for LAS as a function of depth in subsurface soil samples.

As the chart indicates, degradation rates for LAS were rapid in subsurface soil samples collected from beneath the laundromat pond. Degradation half-lives ranged from 2 to 14 days and averaged about 5 days in the majority of the samples tested. The effectiveness of biodegradation as a practical removal mechanism is illustrated in Figure 16, which charts the decrease in measured LAS concentrations in soil strata as a function of depth. Levels of LAS are reduced more than two orders of magnitude, from greater than 200 $\mu g\ g^{-1}$ to less than 2 $\mu g\ g^{-1}$ (the analytical detection limit) over a vertical distance of only 2.5 m. Clearly, therefore, biodegradation is the dominant removal mechanism for LAS in the laundromat pond subsurface environment.

6 LOOKING TO THE FUTURE

Obviously there is a need to characterise biodegradation processes at a more fundamental , molecular level... Procter & Gamble have several programmes directed towards this objective *viz*:

a) determining structure/activity relationships for biodegradability,
b) understanding the detail of the biological processes involved and
c) elucidating the physical, chemical and ecological interactions that occur in surface and subsurface soil systems.

In addition to this programme, which is aimed at characterising biodegradation processes at both cellular and molecular levels, new ways will continue to be developed to follow biodegradation processes in the "real world".

One of these new methods is a recently completed experimental stream facility which is designed to confirm the environmental compatibility of consumer product materials (like soaps and detergents) that enter rivers as part of treated municipal sewage effluent. Several flowing water stream channels can be established in the facility using river water, natural sand and gravel materials and organisms from a local river. Different types of wastewater can be added to the stream channels in order to simulate the conditions that exist in natural rivers that receive treated municipal sewage effluents. The research facility consists of a 450 m^2 building enclosing eight 4 m fibre glass stream channels. Research conducted at the facility will combine the best aspects of both laboratory tests and field experiments including:

a) the use of aquatic plants and animals representative of organisms living in healthy rivers and streams,
b) experimentally controlled, yet realistic environmental conditions (lighting, temperature, flow rates),
c) all year round testing,
d) simultaneous access to various wastewater treatment types and effluent and
e) replication for data reliability.

The facility is largely automated to ensure continuous operation during long-term experiments. A computer controls flow rates of incoming river water and sewage effluents and continuously monitors several water quality parameters (e.g. dissolved oxygen, temperature, dissolved solids, pH etc.).

The experimental stream research facility is a state-of-the-art research tool. The streams are designed to provide key habitat characteristics of natural streams. With this facility, Procter & Gamble expect to develop even more realistic safety assessments in fresh water environments for chemicals of interest and to be able to define more accurately the environment's capacity to handle consumer products, like detergents, that enter it as part of treated municipal sewage.

In conclusion, using the words of an American sage "Biodegradation is simply a matter of faith, hope and acclimation, but the greatest of these is acclimation."

7 ACKNOWLEDGEMENT

The author wishes to thank R.J. Larson and M. Stalmans for helpful discussions during the preparation of this review.

NORSK HYDRO AND ENVIRONMENTAL AUDITING IN THE UNITED KINGDOM: COMMUNICATING GOOD PRACTICE

C. C. Duff

Norsk Hydro (UK) Ltd
Twickenham
Middlesex, TW1 1EE
United kingdom

Norsk Hydro is Norway's largest industrial company, with sales in 1990 of some £5.5 billion, of which over 87% are generated outside Norway. The Group is around the 50th largest in Europe in terms of sales. Roughly 40% of sales are derived from fertilizers of which Hydro is by far the largest producer in the world, 30% from aluminium and magnesium (largest in Europe and largest in the world respectively) and 30% from oil gas and their downstream products, notably PVC polymers.

Hydro employs nearly half a percent of the total population of Norway. The firm is thus a household name, even though it markets few consumer products. Hydro is broadly considered to be synonymous with the chemical industry (fertiliser, plastics and basic industrial chemicals production) as well as being a metals manufacturer. These industries are commonly considered (rightly or wrongly) to be the prime contributors to pollution in industrialised societies.

When an environmental pressure group broke into Hydro's Porsgrunn works in 1988 and publicised the company's findings that soil samples they had taken were contaminated, it was the latter, rather than that the company knew about the problem and was dealing with it , which drew the headlines in the Press. As a result, Hydro decided to undertake a complete environmental audit of all its sites in Norway and to publish the results of its performance against local emission and effluent permits, what was being done in areas of unsatisfactory performance, and what the trends in performance over the previous few years had been. This Miljorapport was published in early 1989. It was circulated very widely, to Hydro employees and their families, and to all other interested parties - local and national authorities, and householders, schools and colleges in the vicinity of all its sites.

The effect was gratifying. Hydro's openness gained the Group new friends and changed people's attitudes. The altered impressions will have contributed to Hydro being voted "the best company to work for" in Norwegian opinion polls over the last year.

In the United Kingdom, Hydro is also a major industrial force. The United Kingdom is Hydro's largest geographical market after Denmark, accounting for over 14% of Group sales. Hydro's United Kingdom based manufacturing companies are number one and two in the United Kingdom markets for fertilisers, PVC, aluminium extrusions and farmed salmon; it has a United Kingdom Oil and Gas exporation company, and new companies manufacturing diecast magnesium bicycle frames and magnesium granules. These companies' sales amounted to £412 million in 1990, and in addition up to £400 million of natural gas was imported into the United Kingdom from the Norwegian sector of the North Sea.

In total this puts Hydro within the top 150 companies in the United Kingdom in terms of sales and yet its size and market standing have been little appreciated. This is probably because the United Kingdom Group has grown mainly through acquisition of existing companies over the last ten years with their own reputations and, again, because there are no consumer products, which gives rise to the belief that commercial opportunities have sometimes been missed as a result. Couple this with environmental awareness moving into the foreground in the United Kingdom, and that the main products have undoubted environmental connotations (fertilisers, plastics and farmed fish) it occurred to the Group's United Kingdom management that an opportunity had arisen to change this image.

The Norwegian Environmental Report was followed by a report covering all Hydro's plants worldwide, and when this was published in the United Kingdom in May 1990, the United Kingdom management announced their plans to produce a comprehensive report about their activities.

There were two fundamental guidelines in planning the report. Firstly, the work would be undertaken in-house, but the findings would be checked by an external, objective authority to forestall any suggestion that only part of the story was being published; the part which flattered the company. Secondly, it was decided at the outset that, whatever the findings, they would be circulated and published widely.

The form of the report was to be that of a compliance audit. Under its "Control of Industrial Major Accident Hazards" regulations, the United Kingdom Department of the Environment requires operations handling potentially explosive or toxic materials to develop and publish detailed procedures for monitoring and for dealing with emergencies. Hydro's two main sites are both covered by these regulations. All the company's sites also handle materials which are covered by the Department of the Environment's Control of Substances Hazardous to Health regulations. Both these sets of rules require detailed records to be kept; the audit was able to draw on these. The main aspects of compliance to be reported on were limits set by the Local Authorities regarding such matters as liquid effluents, emissions to air and noise. In these cases it was possible to draw on existing records, sometimes going back over extended periods, to demonstrate site performance compared with the permitted limits.

The exercise and production of the final report were co-ordinated from Norsk Hydro (UK) Ltd, Hydro's central service company based near London. Norsk Hydro (UK) Ltd provided a format for the report and each subsidiary company then prepared its own review for assemblyinto the final version at the centre. Each company also addressed the matter of environmental impact by describing the uses to which their products were put, their merits compared with those of other products and how they could be recycled or disposed of when they had reached the end of their useful lives in the hands of customers.

At this stage Lloyd's Register Environmental Assurance, the organisation appointed as external assessors, were called in. Their remit was, given the products manufactured at each site and the site's location, to determine which regulations should be observed (appropriate monitoring procedures were instigated and operated to assist in this respect) and to verify that data to be published was accurate and complete. They also inspected the final draft of the report and summarised their findings on all these aspects in an assessor's report which was printed in full in the final version of the report.

What were the results? Lloyd's found that Norsk Hydro complied with statutory emission limits (often by a wide margin) and with monitoring procedures. There were two areas where the company could improve "duty of care". It was found that more could be done to ensure that waste disposal contractors dealt with wastes in a proper manner. In the case of one or two of Hydro's smaller companies, there had been a tendency to assume that no complaint from the authorities meant that the situation must be satisfactory. Companies should always monitor, if only to be able to react quickly if an accidental emission does occur.

The Norsk Hydro "United Kingdom Environmental Report" was launched at a well attended press conference on the 24th October 1990. On the same day "The Financial Times" carried a half-page article about it; which was followed by an advertisement inviting readers to send for copies. Since then Norsk Hydro have had a steady stream of requests for the report; around 700 from 24 countries.

The company sent the report to all United Kingdom employees, to all the Members of Parliament in constituencies where the company had interests, to the Ministries of the Environment, Education & Science and Trade & Industry, to the Pollution Inspectorate, to the European Commission, to environmental pressure groups, to the chairmen of the "Times Top 1000 companies". Norsk Hydro operating companies circulated it to their own local contacts, customers and suppliers. The total sent out so far is over 10,000. The results have surprised even Norsk Hydro's public relations consultants. The Press comment is universally favourable. Two BBC TV programmes have been made (by Tomorrow's World and Business Matters) and requests are continually being received for the company to speak about its experiences, both at conferences and also to individual companies (some of them household names) who are looking for advice on how to prepare their own audits. There is continual contact now with the Directorate General XI in the European Commission, who are seeking opinions about the merits of statutory environmental auditing. The culmination for Norsk Hydro in the United Kingdom has been the appointment of the managing director of Norsk Hydro (UK) Ltd to the advisory committee on business and the environment set up in May 1991 by the Secretaries of State for the Environment and for Trade & Industry.

What next? The first effects of the audit have been developments in Hydro's internal procedures. For example, all major capital expenditure proposals have to include an environmental impact assessment, and internal environmental auditing processes have been refined, whereby all Hydro's major sites around the world are visited successively by teams comprising people from the Group Health, Safety and Environment Department together with operating managers from Hydro sites in other countries. The Group's environmental policy statements have been revised. It is still believed that the company should conduct its own audits. External assessors have a part to play, but it is Hydro who know best what is actually going on at a site, who know best where to find areas for improvement and have the authority and the commitment to bring improvements about. Generally, the message has been reinforced that environmental awareness is a line management responsibility: the chief environmental officer is the chief executive of the operation concerned.

As regards the external world, the first change the Report has introduced is the launch of the "Hydro Award", a £10,000 prize to be awarded for achievement in water quality improvements. Hydro believes it does good work in protecting the environment, so now wants to reward and draw attention to others who are doing notable work in a field which is so relevant to Hydro. Next, as the largest fertiliser producer, the company has a responsibility to provide the full facts about fertilisers and agriculture, so have written and published an authoritative book on the subject, properly referenced and refereed by scientists. A similar book on polymers is now being prepared. Next, dialogue now exists with the National Curriculum Council (which provides guidelines for United Kingdom schools on curriculum content) about using material from the Report as a base for school exercises in science, the environment and industrial and economic understanding. In this field, the coordinator of the Report has been appointed to a Department of Education & Science working party investigating means of introducing the environment more generally and more effectively into college and university education.

The general conclusions from the exercise are that it has been enormously effective. First, Hydro's openness and readiness have been recognised by the authorities and our opinions sought at the formative stage of legislation and regulation setting. Next, in Norway it is felt that environmental (and health and safety) concern on the part of top management are having a positive effect throughout the system, on such matters as productivity and reduced accident levels, as well as cleaner and therefore more cost-effective operations. Hydro expects these effects to be seen in the United Kingdom as well.

As regards the bigger picture, the underlying feature of all this work is *communication*. It is essential to communicate, in this case using the medium of an environmental message, so as to increase the general level of awareness and understanding of industry and its place in society. The barriers of suspicion and mistrust must be broken down if more and better trained young people, especially girls, are to be attracted into the rewarding careers which industry can offer.

THE EEC VIEWPOINT OF THE ENVIRONMENT

A.E. Bennett

Directorate General XI(A)
Commission of the European Communities
Brussels
Belgium

1 INTRODUCTION

Though there is earlier legislation of an environmental nature, the impetus for the creation of the present arsenal of environmental regulations, directives and decisions was given by the Heads of the Member States meeting in Paris in 1972, just a few months after the Stockholm Conference. They recognised that economic expansion was not an end in itself, but that it should "result in an improvement of the quality of life as well as in standards of living and that particular attention should be given to intangible values and to protecting the environment". Consequently they invited the Community institutions to prepare a first Environmental Action Programme.

2 ENVIRONMENTAL ACTION PROGRAMMES

The first programe was adopted in November 1973. It defined the basic principles and objectives of the Community environmental policy, and identified the general actions to be taken over the following years. Its main principles - some of which were subsequently incorporated into the 1987 amendments of the Treaty - are the following:

a) The best environmental policy consists in preventing the creation of pollution and nuisance at the source, rather than trying to remedy the effects subsequently;

b) Effects on the environment should be taken into account at the earliest possible stage in all technical, planning and decision-making processes;

c) Any exploitation of natural resources or of nature which causes significant damage must be avoided, in view of the limited capacity of the natural environments to absorb pollution and to neutralise its harmful effects;

d) The cost of preventing and eliminating nuisances must in principle be borne by the polluter;

e) Care should be taken to ensure that activities carried out in one state do not cause any degradation of the environment in another state.

These principles have been updated and extended in the three subsequent programmes (adopted in 1977, 1983 and 1987), reflecting the evolution of environmental thinking over the last fifteen years, as well as the challenges which have emerged.

The third and fourth programmes emphasised the need to integrate environmental protection requirements into other policies, aiming at what today would be called "sustainable development". The fourth accompanied this with a statement on the need for a more integrated approach to pollution control and reduction, in order to avoid the transfer of pollution from one area to another.

In addition to developing a philosophy of environmental protection and improvement, the programmes set an agenda for specific actions for the years which they covered. For example, the fourth called for harmonised legislation to manage the environmental risks associated with the use and release of genetically modified organisms; now adopted. As concern for the emerging global environmental challenges has increased, so the priority given to the international dimension of the Community environmental policy has grown. The Fourth Action Programme was particularly sensitive to the need to promote international action for the protection of the environment, and in particular the need to assist the developing countries to overcome their special difficulties.

A fifth programme for the years 1993 to 1998 or 2000 is now under discussion prior to presenting a proposal to the Council later this year. It is too early to be able to say what it will contain but it is hoped that the Community's role will be at the global level, treating Europe as a whole at the regional level, where national and local levels will be clearly identified. This would, of course, allow emphasis to be put upon the principle of subsidiarity. In any case, it is clear that policies must now reconcile economic growth with a high level of environmental protection and that this must satisfy the citizen to the extent that they will command public confidence and support.

3 PROGRESS TO DATE

At this point, it is fair to ask what has been achieved. As regards industry a great deal, notwithstanding that there is still much to do. Covered in approximately 300 regulations there are, for example:

a) systems for the classification, packaging and labelling of dangerous substances;

b) the means to control and prohibit dangerous substances and preparations;

c) innovative regulations for the development of biotechnology;

d) and shortly, Community measures for product and process audit;

e) for emissions, an emerging coherence that will shortly lead to an integrated permitting procedure;

f) for waste, a framework of regulations that will govern the pratice of industry in general; and the technical standards and performance of the waste industry itself.

If this seems an onerous load or imposition, one must reflect upon the alternative. This can be seen and experienced outside the European Community without having to travel very far. Furthermore, all the measures are trade supporting or encouraging - the object is not to kill the golden goose but to make sure it can flourish in healthy surroundings.

4 FUTURE STRATEGY

Beginning with waste as this is the principle interest of this symposium. There is a need for a strategic approach based on the broadest notion of integrated pollution control. A three pronged approach can be used:

a) environmental impact assessment - extended to cover a wider range of development and more searching in character. This can be seen as a prerequisite for any form of permit licencing and a system of charging.

b) environmental audit of industrial installations. This is the ongoing control process.

c) product audit and eco-labelling. This is the way by which the design of manufactured goods and use of resources will be influenced.

This outline is in essence a new policy orientation.

Previous environmental policy has been proscriptive in character with an emphasis on the "thou shall not" rather than the "let's work together" approach. The result has been that industry has tended to become preoccupied with questions of compliance rather than in actively seeking to maximise its contribution towards improving environmental quality. In the next decade greater emphasis must be given to providing industry with incentives; environmental considerations must not be seen inevitably as appearing on the debit side of the annual accounts.

In practical terms, factories and other types of industrial installations will have to minimise their negative impact on the environment, in the local, regional and global contexts. This implies strategic choices about the use of energy, water and raw materials. It also requires an analysis of emissions and the possibilities for recycling and saving. Factories will also have to operate at a high level of safety and take all the necessary precautionary measures. To enable this, the Community's regulatory approach in this field is evolving towards a more integrated control of pollution.

Then, and more contentiously, the public should have a more direct role in judging the environmental performance of industrial activities. To this end, companies will need to provide factual and objective information on the environmental characteristics of their activities and on their performance. The basic step is to provide, systematically and periodically, data on pollutants released into the environment from their activities. This would create better awareness within companies themselves of their environmental problems and a sound basis to identify priorities for intervention and competition between companies to achieve emission reductions.

Finally, companies would be encouraged to take a more positive and creative part in environmental protection. Compliance with regulations and standards should not be condsidered sufficient. They should establish their own environmental policy and objectives going beyond compliance, set operational targets and develop management systems to achieve a high level of day-to-day performance. Environmental auditing of these management systems, coupled with information to the public will provide the fundamental tool to improve performance, transparency and *partnership* with the public.

To turn to the problem of products, the increasing demand for consumer goods is resulting in an ever-increasing load on the environment. That load is spread over the whole life cycle from manufacture to disposal.

As society is becoming more aware of the environmental problems, consumers are demanding more information on the environmental qualities of the products which they purchase. Environmental policy has to ensure that the right mechanisms are put in place to enable societies to increase their consumption of products in line with their expectations whilst reducing the overall impact on the environment. This implies:

a) that products are designed in such a way that their production and use consume a minimum of natural resources and energy and that their disposal minimises the quantity of materials that cannot be reused in one form or another.

b) that the consumer is given proper information on their use and the provisions for their disposal.

The eco-labelling scheme will be a direct encouragement for manufacturers to produce environmentally friendly products. The award of an "Eco-Label" will be a clear message to consumers that a given product is superior to its competitors with regard to environmental impact. Perhaps of greater importance is the opportunity it will give to the individual consumer to have a direct influence on environmental quality as a result of day-to-day purchasing decisions.

5 LEGISLATION ON CHEMICAL SUBSTANCES

In 1967 the Council adopted a first Directive in order to unify the laws in the Member States relating to the classification, packaging and labelling of dangerous substances. This Directive has been repeatedly updated over the years and the 6th amendment, adopted in 1979, is the version which is in force today. It has two aims:

a) protection of man and the environment and

b) elimination of technical barriers and the improvement of trade.

The object of classification is to identify all the physico-chemical and toxicological properties of the substance which may constitute a risk during normal handling or use. The Directive lists fourteen categories in which dangerous substances may be classified. The labelling of a substance is derived from this classification and the most severe hazards are highlighted by symbols. The dangerous properties of the substance are specified in standard risk phrases and safety phrases giving advice on the necessary precautions. The label must take account of all potential hazards which are likely to occur during normal handling or use. It must draw the attention of people handling dangerous substances to the dangers involved and must clearly show:

a) the name of the substance
b) the symbol and the indication of danger involved
c) standard phrases indicating the nature of special risks.
d) the name and address of the manufacturer or the importer.

The Directive has a number of technical annexes. Annex 1, which is regularly updated, lists all the substances for which uniform labelling in the Community has been agreed. In practice, a manufacturer or an importer into the European Community of a dangerous chemical substance has to label that substance before he puts it on the market. If the substance is listed in Annex 1 of the Directive, the label contained in that Annex must be applied.

In the case where the substance is not yet listed in Annex 1, it must be provisionally labelled by the manufacturer or the importer. This provisional label ramains valid until the substance is listed in Annex 1 thus signifying that a label has been fixed at Community level.

<u>The inventory of existing substances</u>

The Commission, in collaboration with the Member States and

industry, has established an inventory of substances marketed in the European Community. These substances are the so-called "existing chemical substances" on the Community market before 18 September 1981.

The European Inventory of Existing Commercial Chemical Substances (EINECS) contains about 100 000 chemical substances which are on the Community market. Industry estimates that about 20 000 of these substances on the Community market are dangerous in the sense of the 6th Amendment Directive: 19 000 have still to be agreed upon, and added to Annex 1. Priority is being given to:

a) substances which have carcinogenic, mutagenic, or teratogenic properties;

b) substances for which the provisional labelling is different between companies;

c) particular substances which Member States request the Commission to consider adding to Annex 1.

Currently 500 pesticides, 400 solvents and 100 other substances are under examination together with 170 suspected carcinogens.

New substances

The 100 000 existing chemical substances listed in the EINECS were on the Community market before September 1981. By definition all substances which are not included in this inventory must be considered as new substances. For these new substances the Directive foresees that the substance must be tested prior to being marketed and a notification dossier must be submitted to the Commission. The procedure for new substances has been in force for six years; a total of about 700 substances has been notified. This number is increasing and is expected to average about 300 notification per year.

Classification and labelling of dangerous preparations

Most of the chemical products on the market are mixtures of more than one substance and must be considered as preparations. The Council has recently adopted a general Directive on the labelling of preparations - the principles and definitions of which are the same as in the Directive for dangerous substances.

Directive on prohibition and restriction of marketing of certain substances

There are some substances which are so dangerous that restrictive measure have to be taken. To this end, several Directives have been adopted banning or severely restricting the use of named substances; many of these are pesticides and the remainder are industrial chemicals.

Export notification

Many industrial countries have developed lists of dangerous substances which are banned or severely restricted at national level. Nevertheless, all too often these continue to be produced for export to third countries. In June 1988 a regulation was adopted concerning export from and import into the Community of such dangerous chemicals. This lays down a notification procedure whereby the importing country is informed of the regulatory restrictions in the Community and the reasons for them.

6 INTEGRATED CHEMICALS CONTROL

A systematic approach to chemicals control should be built on the sequence of data collection, risk assessment and risk reduction with the necessary control measures. However, the only systematic collection of data on chemicals is the new notifications scheme established under the 6th Amendment to the Directive on classification, packaging and labelling. For existing chemicals there is as yet no systematic collection of data; instead data collection is initiated for "priority chemicals" identified for various targets and environmental compartments, often in the context of a particular

Community Directive, for example, water pollution, waste disposal, atmospheric pollution and worker protection.

Risk reduction measures are often aimed in a rather *ad hoc* manner towards those chemicals which are the subject of the latest scare stories in one or more Member States. By the time control measures come into force, the chemicals in question may no longer be produced in the Community. Furthermore, the same chemical may be treated inconsistently under different Community instruments.

In recognition of the need to achieve a more integrated approach to chemicals control, the Commission has recently put forward two measures which should go some considerable way to a more integrated approach:

a) In the 7th amendment to Directive 67/548/EEC on the classification, packaging and labelling of dangerous substances Article 1 has been amended to include risk assessment of new (notified) chemicals as being one of the objectives of the Directive. Article 26 of the Directive also foresees a mechanism for agreeing appropriate risk management options for chemicals in the context of the Directive. Thus, the Directive will cover data collection, assessment and recommendations for risk management.

In view of the changes foreseen for the 7th amendment, considerable progress has already been made in discussion with the Member States towards the elaboration of common principles for the assessment of new chemical substances. These common principles will, to a large degree also be applicable to existing chemical substances. In essence the main thrust of this work is to agree on mechanisms for identifying substances as potential candidates for risk managment measures, for example, emission controls, restrictions on marketing and use etc.

b) Proposed regulation for the systematic evaluation of existing chemicals foresees the collection of available data on existing chemical substances and, on the basis of these data, the identification of priorities for more in depth risk assessment. This assessment will subsequently lead to recommendations for risk management actions.

This regulation will provide a unique opportunity to implement a coherent and integrated approach to dealing with the 100 000 existing chemical substances on the Community market. Furthermore, it will question the necessity of developing specific data collection and assessment measures, including priority lists, in relation to individual Community regulatory instruments. Instead the specific concerns related to target groups of environmental media should be incorporated into the overall process of priority setting and risk assessment.

7 CONCLUSION

The thrust of these remarks is towards the need for more integrated pollution control and for improved chemicals control. A more integrated approach demands better cooperation. To this end the most exciting development so far has been the emergence of responsible care programmes. Pioneered by the chemical industry, responsible care programmes go beyond what can ever be achieved by conflict and restriction, or even by punitive costs imposed through economic measures. Responsible care is a partnership approach which is the way forward for the future.

HER MAJESTY'S INSPECTORATE OF POLLUTION'S ROLE IN REGULATING INDUSTRIAL RELEASES TO THE ENVIRONMENT

D. Slater

Her Majesty's Inspectorate of Pollution
Romney House
43 Marsham Street
London, SW1P 3PY
United Kingdom

1 INTRODUCTION

We are living in an age of unparalleled public interest in environmental issues and that is to be welcomed. Public concern has been running at a consistently high level for the past two to three years and the Government has responded with a series of important measures aimed at preventing pollution and protecting our environment.

Three particularly substantial environmental initiatives came to fruition recently: in September 1990 a White Paper on the Environment was published; in November the Environmental Protection Act received Royal Assent. The centre piece of the Act as far as industry is concerned is our new system of Integrated Pollution Control (IPC); and in July the Prime Minister announced the setting up of an Environmental Agency.

The Government, industry and individuals alike must face up to the major environmental challenges that are now with us and will continue to arise in the future. It is clear that industry is beginning to meet its side of the challenge by participating actively in moves to "green" its practices. There is little doubt that a sound environmental strategy will improve a company's local, national and even international reputation. During the 1960's a lot of time and energy was wasted in a rather futile debate. Some environmentalists saw industrial advance as a harbinger of destruction. Industrialists responded by attacking its critics as wishing to lead us back to agrarian poverty. The situtation has since changed and now industry can win praise for its environmental achievements. Environmentalists are now working alongside industrialists to improve the quality of our environment.

Alongside industry's own efforts, however, there is also an important place for Government regulation. A judicious mix of private sector initiative and regulation is seen as the way forward towards higher technological standards and better performance in a more environmentally demanding market. The implementation of Part 1 of the Environmental Protection Act 1990 will make a major contribution here. The Act sets a clear framework of pollution control requirements with which industry can comply and which the public can identify as positive steps towards a cleaner environment.

In the past pollution control systems in the United Kingdom have developed in response to the needs of one of the three environmental media of air, land and water. But it really does not make sense to treat the release of wastes from major processes to one of the environmental media without considering the other two. With IPC a multi-media approach has been developed. In this the United Kingdom is among the leaders in Europe and the rest of the world. Indeed the European Commission is currently developing an instrument to introduce a system of IPC Community-wide. It will be good for British industry to be ahead of the field in having systems suited to the IPC approach.

2 HER MAJESTY'S INSPECTORATE OF POLLUTION'S ROLE

Her Majesty's Inspectorate of Pollution's (HMIP) primary functions are to implement the provisions of several major pieces of legislation - not only under the Environmental Protection Act but also under the Radioactive Substances Act 1960, the Water Act 1989, the Alkali Act 1906 and the Health and Safety at Work Act 1974. The current functions of HMIP can be summarised as follows:-

a) provision of authoritative and independent advice to Government on pollution control practices;

b) approval, inspection and oversight of potentially polluting processes under the Alkali Act, Water Act and Health & Safety at Work Act - currently some 3,000 installations;

c) registration/authorisation of premises both holding and disposing of radioactive materials under the Radioactive Substances Act - a further 9,000 premises;

d) regulatory oversight and audit of waste regulation authorities;

e) research on the regulatory aspects of pollution; and

f) implementing the relevant sections of Part 1 of the Environmental Protection Act 1990 dealing with Integrated Pollution Control - around 5,000 installations.

Separate but similar arrangements apply in Scotland and Northern Ireland.

The author's role as the Chief Inspector of HMIP is to ensure that these regulatory responsibilities are carried out to guidelines and standards laid down by the Secretary of State for the Environment in England and the Secretary of State for Wales and as Director of HMIP is to lead the Inspectorate into a position where it operates as a regulatory agency that is fully in tune with modern demands. This means a more structured relationship with industry, greater independence from central Government and increased accountability to members of the public. To do this new ideas are needed, for example, the introduction of first time cost recovery charging to recoup HMIP's costs. An extensive public registry system has been set up to provide access to a wide range of environmental information. With greater public accountability the Inspectorate also has a duty to inform the public and industry about HMIP's aims and responsibilities.

3 IMPLEMENTATION OF IPC

IPC will require that no prescribed process can be operated without a prior authorisation from HMIP after a date specified in regulations. The processes and substances to be controlled under IPC and the timetable for their introduction into the new system are set out in detail in the Environmental Protection (Prescribed Processes and Substances) Regulations 1991; a summary is provided in the appendix.

The Environmental Protection (Applications, Appeals and Registers) Regulations 1991 determine the procedures for applying to HMIP for an authorisation, the information required by HMIP, the bodies which HMIP must consult and requirements for advertising the application and for placing relevant information in a public register. The requirements for involving the public in the process are a key aspect of IPC. They reflect the philosophy that the public has a right to know about pollution issues (subject to safeguards where essential for commercial confidentiality). HMIP is required either to grant an authorisation, subject to any conditions which the Act requires or empowers it to impose, or to refuse it. HMIP must refuse it unless it is considered that the applicant will be able to carry on the process in compliance with the conditions to be included in the authorisation. The application procedure is illustrated in Figure 1.

In setting the conditions within an authorisation, Section 7 of the Act places HNIP under a duty to ensure that certain objectives are met. The conditions should ensure that:

Figure 1 Integrated pollution control procedure.

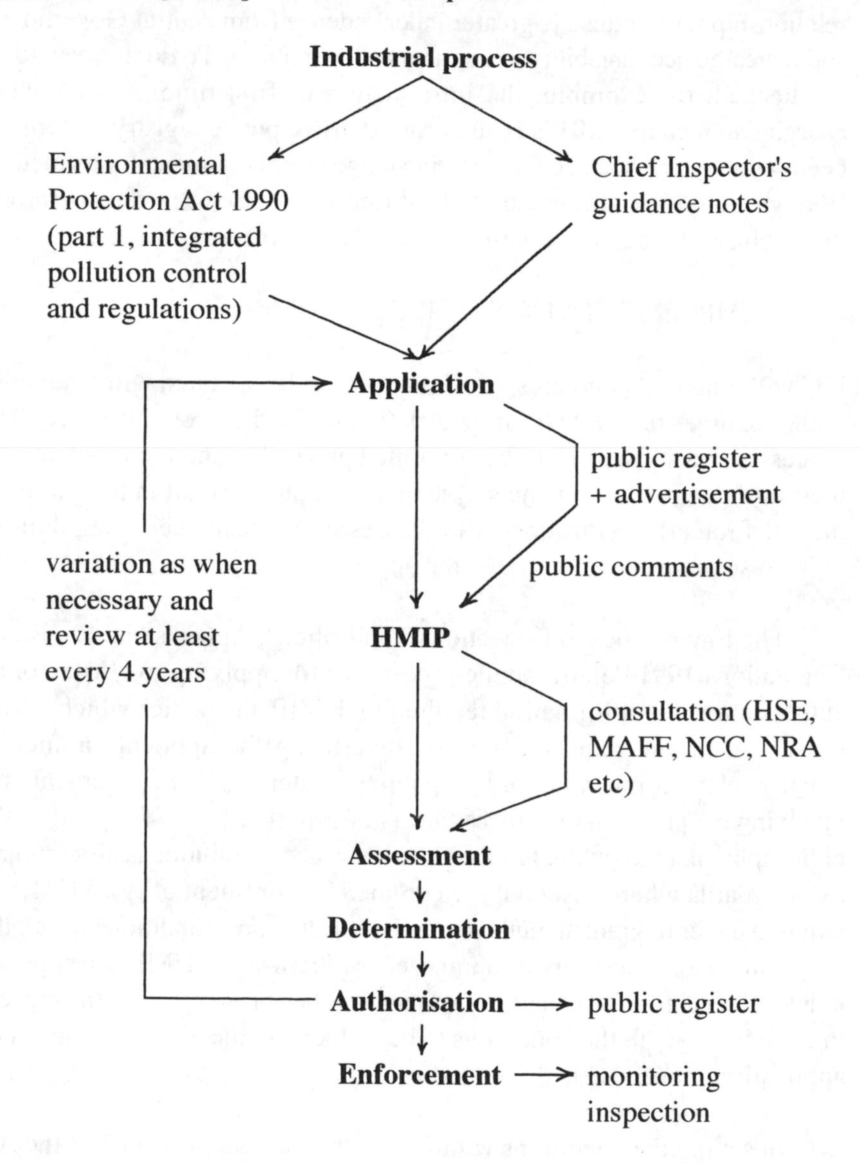

a) the best available techniques (both technology and operating practices) not entailing excessive cost (BATNEEC) are used to prevent or, if that is not practicable, to minimise the release of prescribed substances into the medium for which they are prescribed; and to render harmless both any prescribed substances which are released and any other substance which might cause harm if released into any environmental medium;

b) releases do not cause, or contribute to, the breach of any direction given by the Secretary of State to implement European Community or internatonal obligations relating to environmental protection, or any statutory environmental quality standards or objectives, or other statutory limits or requirements;

c) when a process is likely to involve releases into more than one medium (which will probably be the case in many processes prescribed for IPC), the best practicable environmental option (BPEO) is achieved (i.e. the releases from the process are controlled through the use of BATNEEC so as to have the least effect on the environment as a whole).

The concept of BATNEEC contains an inbuilt dynamic towards higher standards because as the technology available improves environmental protection standards will be raised.

Process operators and indeed the public will require an assurance that BATNEEC is applied in a rational and consistent way. BATNEEC standards for each class of IPC process will be set out in published guidance notes. In preparing the guidance notes HMIP will review available techniques internationally as well as tapping industry's own expertise and experience. At the outset, industry, through its various representative bodies, will have an opportunity to offer views on the factors that will need to be covered in each note. Before a note is finalised it will be issued in draft for comment and discussion by all interested parties.

4 EXPERIENCE TO DATE

Although IPC has only been in being since 1st April 1991, there are already some early indications of its impact on industry and the public. The first tranche of processes to come under IPC included, of necessity, large combustion processes, mostly power stations. A number of applications for confidentiality of information followed. These were swiftly rejected by HMIP and appeals have been lodged with the Secretary of State. A clear message was sent to industry - that the new system is to be an open system. The results of the appeals are awaited with interest.

As with all innovatory and complex new systems, the implementation of IPC has involved tackling a number of problems quickly and decisively. Early indications show that the application of logging and tracking systems can work well given the necessary administrative support. There has been a major problem in that applicants did not provide sufficient information to enable HMIP to carry out the environmental assessments necessary to determine IPC authorisations. At the end of July applicants were informed of the need to provide further information and field inspectors visited each applicant to discuss information requirements. This meant extending the time period for determining the applications from two to six months. The message to industry was clear: HMIP is determined to get the system right first time and that industry must provide the information necessary for HMIP to do its job properly.

A great deal of interest has been expressed in the public registers; it is intended to improve this service. It is too early yet to give judgement on other aspects of the new system but the signs are very encouraging.

5 HMIP'S RELATIONSHIP WITH INDUSTRY

With the introduction of IPC there is a more structured relationship between the Inspectorate and industry. Industry must be the provider of information on which HMIP determines authorisations - a point emphasised by HMIP going back to industry to request more information. Industry must also demonstrate compliance with authorisations by carrying out monitoring and instituting quality assurance procedures etc.

HMIP will, of course, carry out inspections and conduct independent monitoring surveys, in future HMIP's inspections will be more intensive than in the past and will be looking not only for compliance with authorisations but also to ensure that the"industrial techniques" used are of the required standard, i.e. consistent with BATNEEC. This includes quality assurance, training and related aspects.

Industry must not be squeezed to death by paper mountains and requests for infinite detail about processes. There is a balance to be struck so that industry can operate efficiently and effectively using clean technologies to ensure that no harm is done to the environment. HMIP in turn must be equally efficient in determining applications and in enforcing authorisations. The systems must be publically accountable but should not be a hindance to environmentally responsible companies.

6 HMIP'S RELATIONSHIP WITH CITIZENS

Along with the unprecedented concern among members of the public about environmental issues, the public is becoming increasingly knowledgeable about the environment and how it works. This represents perhaps the most significant change over recent times. Recognising this the Government has responded by publishing the recent Environment White Paper and the anniversary report, and by bringing in new legislation such as the EPA90. This was a clear recognition that the environment is everybody's responsibility and that the public is an interested and active player. The way the public buys goods, moves about, respects its own background and votes, is now a crucial and influential factor in the partnership needed for effective environmental protection. As far as interaction with HMIP is concerned, the public must be seen as an additional arm to the regulatory arsenal. When a member of the public, either as an individual or as a member of a pressure group, suspects that a problem exists in the environment then HMIP must welcome being alerted to suspend pollution. Visitors are welcome to HMIP's regional offices to find out more about applications for IPC and other authorisations; particular attention will be paid to this service as HMIP intends to be user friendly to members of the public. Over the coming months ways of expanding the service to the public will be explored as part of the culture change towards an inspectorate which is fully in tune with modern requirements. The public as "green watchdogs" can help HMIP keep on its toes in carrying out its job of regulating industry. HMIP must in turn move towards a closer, more responsive relationship with members of the public and with responsible and respected pressure groups.

7 THE FUTURE

Finally, a brief mention must be made of the Government's decision to establish a new environmental agency by amalgamating HMIP with the relevant parts of the National Rivers Authority (NRA) and other similar bodies. In his speech on the global environment recently, the Prime Minister praised the work of existing environmental agencies such as HMIP and NRA. He went on to say that:

> "There is scope to achieve even better integration between the different agencies, as foreshadowed in last year's White Paper on the Environment. I can announce today that we plan to set up an Environment Agency. This will bring together Her Majesty's Inspectorate of Pollution and related functions of the National Rivers Authority to create a new agency for environmental protection and enhancement. It would have responsibilities for monitoring the state of air and water. It would make proposals to Government on standards. It would regulate emissions and discharges to achieve the standards set by Government. Again, as foreshadowed in the White Paper, we are also minded to give the new agency responsibility for regulating, handling and disposal of waste, which is currently under the combined supervision of local authorities and HMIP. It is right that the integrity and indivisibility of the environment should now be

reflected in a unified agency. I am confident that this will be a significant step forward."

The decision is good news and adds fresh impetus to the development of a truly integrated environmental regulation agency covering all aspects of industrial pollution control. HMIP is at the very centre of this development and will be working hard over the coming months using experience gained in amalgamating several inspectorates in 1987 to create an environmental agency that is the envy of Europe.

APPENDIX 1

TIMETABLE FOR IMPLEMENTING INTEGRATED POLLUTION CONTROL

EPA schedule 1 reference	Process	Comes within IPC	Apply between	Chief Inspector's guidance note issued
	Fuel & power industry			
1.3	combustion (>50 MWh)	1/4/91	1/4/91 & 30/4/91	1/4/91
	boilers and furnaces			
1.1	gasification	1/4/92	1/4/92 & 30/6/92	1/10/91
1.2	carbonisation	1/4/92	1/4/92 & 30/6/92	1/10/91
1.3	combustion (remainder)	1/4/92	1/4/92 & 30/6/92	1/10/91
1.4	petroleum	1/4/92	1/4/92 & 30/6/92	1/10/91
	Waste disposal industry			
5.1	incineration	1/8/92	1/8/92 & 31/10/92	1/2/92
5.2	chemical recovery	1/8/92	1/8/92 & 31/10/92	1/2/92
5.3	waste derived fuel	1/8/92	1/8/92 & 31/10/92	1/2/92
	Mineral industry			
3.1	cement	1/12/92	1/12/92 & 28/2/93	1/6/92
3.2	asbestos	1/12/92	1/12/92 & 28/2/93	1/6/92

EPA schedule 1 reference	Process	Comes within IPC	Apply between	Chief Inspector's guidance note issued
3.3	fibre	1/12/92	1/12/92 & 28/2/93	1/6/92
3.5	glass	1/12/92	1/12/92 & 28/2/93	1/6/92
3.6	ceramic	1/12/92	1/12/92 & 28/2/93	1/6/92
	Chemical industry			
4.1	petrochemical	1/5/93	1/5/93 & 31/7/93	1/11/92
4.2	organic	1/5/93	1/5/93 & 31/7/93	1/11/92
4.7	chemical pesticide	1/5/93	1/5/93 & 31/7/93	1/11/92
4.8	pharmaceutical	1/5/93	1/5/93 & 31/7/93	1/11/92
4.3	acid manufacturing	1/11/93	1/11/93 & 31/1/94	1/5/93
4.4	halogen	1/11/93	1/11/93 & 31/1/94	1/5/93
4.6	chemical fertiliser	1/11/93	1/11/93 & 31/1/94	1/5/93
4.9	bulk chemical storage	1/11/93	1/11/93 & 31/1/94	1/5/93
4.5	inorganic chemical	1/5/94	1/5/94 & 31/7/94	1/11/93
	Metal industry			
2.1	iron & steel	1/1/95	1/1/95 & 31/3/95	1/7/94
2.3	smelting	1/1/95	1/1/95 & 31/3/95	1/7/94
2.2	non-ferrous	1/5/95	1/5/95 & 31/7/95	1/11/94

EPA schedule 1 reference	Process	Comes within IPC	Apply between	Chief Inspector's guidance note issued
	Other industry			
6.1	paper manufacturing	1/11/95	1/11/95 & 31/1/96	1/5/95
6.2	di-isocyanate	1/11/95	1/11/95 & 31/1/96	1/5/95
6.3	tar & bitumen	1/11/95	1/11/95 & 31/1/96	1/5/95
6.4	uranium	1/11/95	1/11/95 & 31/1/96	1/5/95
6.5	coating	1/11/95	1/11/95 & 31/1/96	1/5/95
6.6	coating manufacturing	1/11/95	1/11/95 & 31/1/96	1/5/95
6.7	timber	1/11/95	1/11/95 & 31/1/96	1/5/95
6.9	animal & plant treatment	1/11/95	1/11/95 & 31/1/96	1/5/95

APPENDIX 2

PRESCRIBED SUBSTANCES

Release to air

Oxides of sulphur and other sulphur compounds
Oxides of nitrogen and other nitrogen compounds
Oxides of carbon
Organic compounds and partial oxidation products
Metals, metalloids and their compounds
Asbestos (suspended particulate matter and fibres), glass fibres and mineral fibres
Halogens and their compounds
Phosphorus and its compounds
Particulate matter

Release to water

Mercury and its compounds
Cadmium and its compounds
All isomers of hexachlorocyclohexane
All isomers of DDT
Pentachlorophenol and its compounds
Hexachlorobenzene
Hexachlorobutadiene
Aldrin
Dicldrin
Endrin
Polychlorinated biphenyls
Dichlorvos
1,2-Dichloroethane
All isomers of trichlorobenzene
Atrazine
Simazine

Tributyltin compounds
Triphenyltin compounds
Trifluralin
Fenitrothion
Azinphos-methyl
Malathion
Endosulfan

Release to land

Organic solvents
Azides
Halogens and their covalent compounds
Metal carbonyls
Organometallic compounds
Oxidising agents
Polychlorinated dibenzofuran and any congener thereof
Polychlorinated dibenzo-*p*-dioxin and any other congener thereof
Polyhalogenated biphenyls, terphenyls and naphthalenes
Phosphorus
Pesticides, that is to say, any chemical substance or preparation prepared or used for destroying any pest, including those used for protecting plants or wood or other plant products from harmful organisms; regulating growth of plants; giving protection against harmful or unwanted effects on water systems, buildings or other structures, or on manufactured products; or protecting animals against ectoparasites.
Alkali metals and their oxides and alkaline earth metals and their oxides.

ROLE OF THE NATIONAL RIVERS AUTHORITY IN WATER POLLUTION CONTROL

A. M. C. Edwards

National Rivers Authority
Yorkshire Region
Leeds
West Yorkshire, LS1 2QG
United Kingdom

1 INTRODUCTION

Rivers carry excess rainfall from the land to the sea. In so doing they provide a habitat for wildlife and a resource used extensively by man. These uses include power generation, transport, water for agriculture, industry and public supply, fisheries, recreation and an indispensable receptacle for wastewaters. The prevention of pollution resulting from the overburdening of rivers with effluent or from intermittent pollution incidents is of prime importance for protecting and enhancing the quality of this resource. This paper outlines the approach to water quality management in England and Wales using examples from the rivers of the Yorkshire Region.

The Water Act 1989 established the National Rivers Authority (NRA), which is vested with the responsibility for combating pollution in "controlled waters". These waters are rivers, canals, lakes (which form an integral part of river systems), estuaries, coastal waters (within the three mile limit) and underground waters.

The NRA's aims related to water quality are:

a) achieve a continuing improvement in the quality of rivers, estuaries and coastal water through the control of water pollution.

b) assess, manage, plan and conserve water resources and to maintain and improve the quality of water for all those who use it.

c) maintain, improve and develop fisheries.

d) conserve and enhance wildlife, landscape and archaeological features associated with waters under NRA control.

e) ensure that dischargers pay the costs of the consequences of their discharges and, as far as possible, recover the costs of water environment improvements from those who benefit.

f) improve public understanding of the water environment and the NRA's work.

The Water Act 1989 provides the duties and powers for the control of water pollution. In brief, it is an offence "to cause or knowingly permit any poisonous, noxious or polluting matter or any solid waste matter to enter any controlled water". The NRA has powers to prosecute for such pollution subject to specified defences, for example where a discharge meets its consent limits (waste water permits - see section 3) or where a discharge is made in an emergency so as to avoid damage to life or health.

Other duties on the NRA under the Act are to monitor the extent of pollution in controlled waters, ensure the achievement of water quality objectives, determine applications for discharge consent and maintain public registers of consents and effluent and water quality data.

2 WATER QUALITY OBJECTIVES

The framework for water quality management in the United Kingdom is based on setting a water quality objective (WQO) for each reach of river and estuary and for coastal and underground waters. These objectives are in the form of quality classifications which are defined by quantitative environmental quality standards (EQS). These classifications and standards are related to the use made of the watercourse, or its potential use if it is currently polluted. These uses include fisheries, fitness for abstraction, wildlife, water recreation and amenity.

As an example, Figure 1 shows the rivers of the NRA's Yorkshire Region. This has an area of 13,500 km^2, a population of 4.5 million people and is principally the catchment of the Yorkshire Ouse which flows into the Humber Estuary. It shows the classification scheme of the former National Water Council.[1] The principal rivers of the northern part of the region are mostly of high quality and are extensively used as a source of water for public supply and agriculture (and to a lesser extent manufacturing industry). They support valuble fisheries and are of high amenity value. The aim is to protect this high quality and enhance it where practicable.

The densely populated and industrialised catchments of the Rivers Aire, Calder, Don and Rother have experienced well over a century of serious water pollution. The aim is to remove this pollution, but a virtually pristine quality (Class 1) is unpractical. A quality objective of "Fair" (Class 2) was set for these rivers by the former Yorkshire Water Authority in 1979 after extensive consultation. Table 1 compares the state of the Region's freshwater rivers and canals in 1990 with the "inherited" objectives. It can be seen that 882 km (15%) of the length of freshwaters are seriously polluted (Classes 3 and 4).

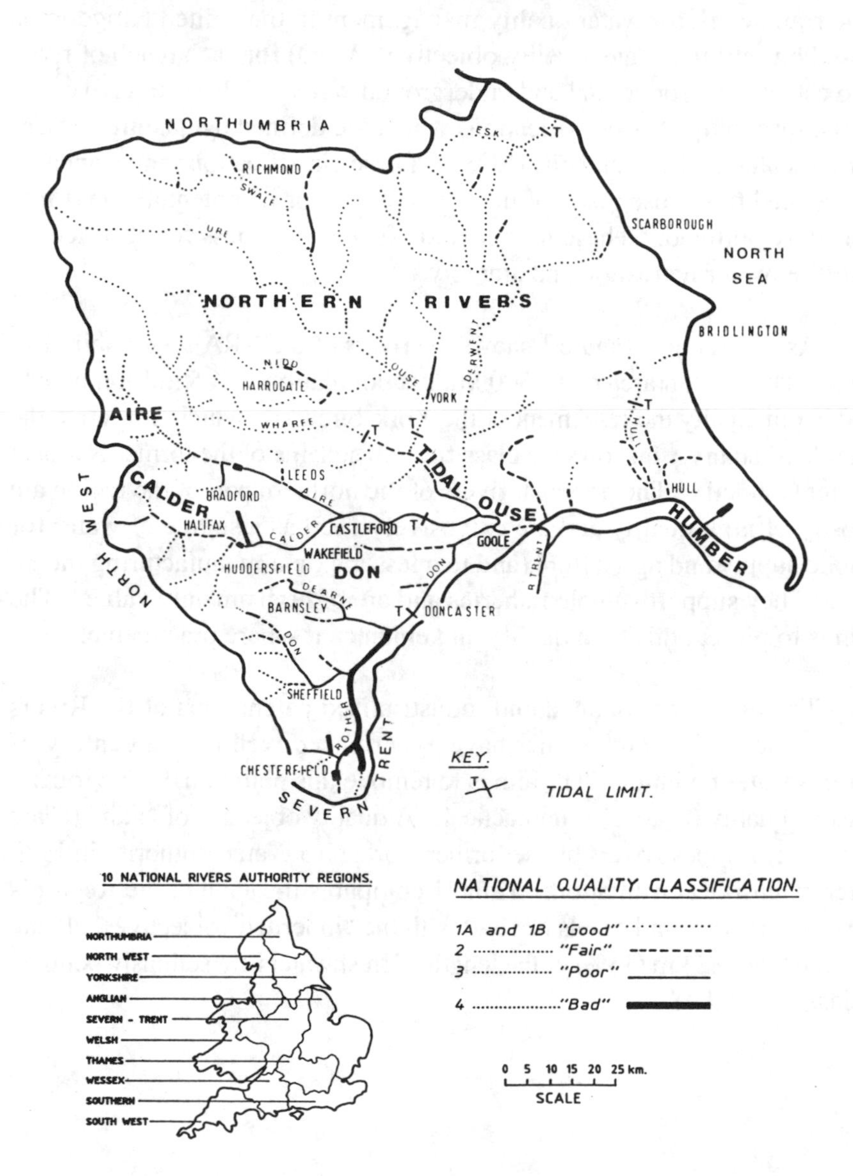
Fig.1
RIVER QUALITY IN YORKSHIRE 1990
NORTHUMBRIA
NORTHERN RIVERS
NORTH SEA
NORTH WEST
AIRE
CALDER
DON
TIDAL OUSE
HUMBER
SEVERN TRENT
ESK
RICHMOND
SWALE
URE
SCARBOROUGH
BRIDLINGTON
NIDD
HARROGATE
OUSE
YORK
DERWENT
WHARFE
HULL
LEEDS
BRADFORD
HALIFAX
CASTLEFORD
WAKEFIELD
GOOLE
HUDDERSFIELD
DEARNE
BARNSLEY
DONCASTER
SHEFFIELD
ROTHER
CHESTERFIELD
TRENT
KEY.
TIDAL LIMIT.
10 NATIONAL RIVERS AUTHORITY REGIONS.
NORTHUMBRIA
NORTH WEST
YORKSHIRE
ANGLIAN
SEVERN - TRENT
WELSH
THAMES
WESSEX
SOUTHERN
SOUTH WEST
NATIONAL QUALITY CLASSIFICATION.
1A and 1B "Good"
2 "Fair"
3 "Poor"
4 "Bad"
0 5 10 15 20 25 km.
SCALE

Table 1 Quality Classification of Yorkshire's Freshwater Rivers and Canals.

National Class	Quality		1989 /km (%)	Inherited Quality Objectives/km (%)
1A	good		2255 (37.4)	2389 (39.6)
1B	good		1982 (32.8)	2467 (40.9)
2	fair		915 (15.2)	1149 (19.0)
3	poor		708 (11.7)	29 (0.5)
4	bad		174 (2.9)	0 (0.0)
		Total	6034 (100)	6034 (100)

The inherited and informal objectives set by all of the former ten regional water authorities are to be replaced by statutory WQO's under the 1989 Act. The Secretary of State for the Environment has the duty to issue classifications to describe water quality and has the task of setting the WQO's based on advice from the NRA.

Once the New WQO's are in force, it will be a duty on both the Secretary of State and the NRA to ensure, "so far as it is practical", that they are achieved; timetables will be set where the standards cannot be met immediately. Objectives may be reviewed by the Secretary of State at five yearly intervals if the NRA requests a review. It is expected that the first statutory WQO's will be set in 1992 after public consultation; in time they will cover all controlled waters.

Increasingly environmental quality standards are being laid down by European Community Directives. Those which are most influential for water quality are given in Appendix 1. The most important Directive for EQS's is the Directive concerning "certain Dangerous Substances discharged into the Aquatic Environment of the Community" and its growing number of daughter Directives. The Directive provides for a List 1 (black list) of the most dangerous substances - those which are toxic, persistent in the environment and accumulated by organisms. The List II (grey list) covers rather less dangerous contaminants with standards being set by national governments rather than in Directives. Appendix 2 lists the substances for which EQS's are in force or have been published.

The speed of passing daughter directives to set conditions for the 129 substances originally identified for List I has been slow. The United Kingdom Government has thus introduced a "red list' related to Integrated Pollution Control (section 3.3) to accelerate the process and to meet the obligations for the environmental protection of the North Sea. Agg and Zabel[2] outline the procedures used for the selection of "red list" substances. Those identified to date are given in Appendix 2.

The NRA has comprehensive programmes monitoring the chemical, biological and microbiological quality of controlled waters. Figure 2 shows dieldrin and hexachlorocyclohexane levels in the Rivers Aire and Calder. The EQS given in the European Community Directive is in the form of an annual average, so the graphs show rolling twelve monthly averages. In the mid 1980's the EQS's were exceeded because of the residues of pesticides derived from sheep dips, which were washed off the raw wool used in the West Yorkshire textile industry, or from the mothproofing of cloth. The replacement of organochlorine compounds with ones which are less persistent has resulted in the decrease of the concentrations found in the rivers so that the EQS's are now met. It should be noted that pesticides were in trade effluents discharged to sewers which then passed through sewage treatment works into the rivers.

3 THE CONTROL OF EFFLUENT DISCHARGES

3.1 Discharge Consents

A discharge of sewage or liquid matter (other than uncontaminated surface water) from trade premises into controlled waters is unlawful unless it is authorised by consent issued under the Water Act 1989. Trade discharges include effluents from agricultural premises, fish farms, premises used for scientific research or experiment, quarries, mining sites and

FIGURE 2.

COMPARISON OF 12 MONTHLY MEAN CONCENTRATIONS OF LINDANE & DIELDRIN WITH THE ENVIRONMENTAL QUALITY STANDARD (EQS).

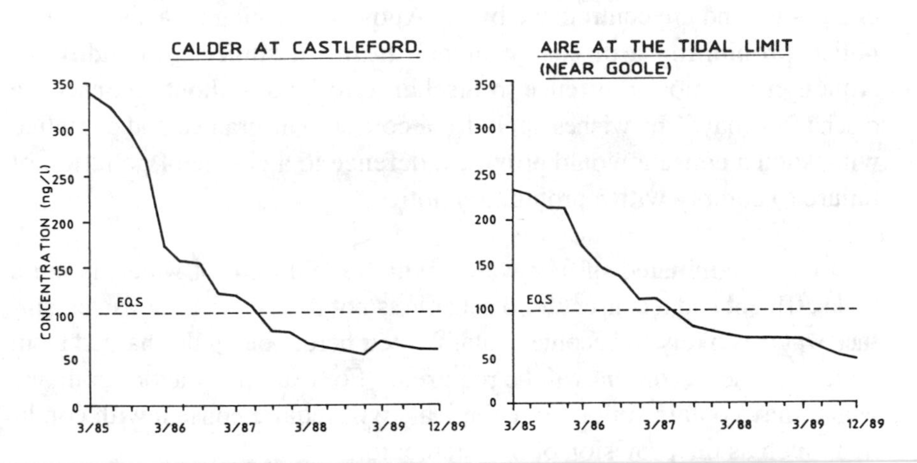

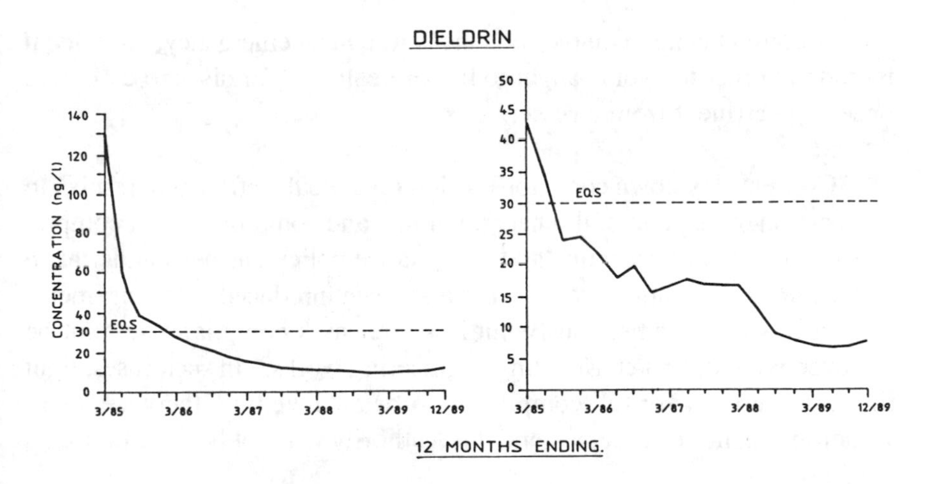

opencast mining spoil heaps. Minewaters from abandoned mines are exempt from control. Such discharges in the older parts of the Yorkshire coalfield cause serious pollution by the precipitation of iron oxides blanketing the river bed with ocherous deposits.

A discharge consent is not required where a discharge of trade or sewage effluent is made to land unless the matter directly enters a stream. Discharges to land are controllable by the Authority serving on a discharger a notice prohibiting a discharge or permitting it subject to conditions. Although it is not an offence to discharge to land without a consent, a discharger may if he wishes apply for a consent. If granted and complied with, such a consent would provide a defence to a charge of pollution or failure to comply with a prohibition notice.

Uncontaminated surface water (which includes roof water) is not a trade effluent and a consent for discharge is not required. If, however, surface water is likely to become contaminated before being discharged from trade premises a consent will be required. For example: factory drainage which may contain spilt oil or chemicals will require consent with conditions such as the provision of oil interceptors.

A consent is not required for a discharge in an emergency, provided it is made in order to avoid danger to life or health, or for discharge of trade or sewage effluent from a vessel.

Consents lay down conditions which regulate the effluent in respect to the location of its point of discharge, volume and composition. A comprehensive review of consenting and compliance policy has been undertaken by the NRA.[3] A charging scheme has also been inroduced.[4] Infringements of consents are taken seriously but in some cases compliance cannot be achieved without first constructing engineering works. In such cases, tight timetables are set for full compliance to be achieved. Otherwise, clear violations cannot be accepted and the Authority will not hesitate to use its legal powers.

3.2 Setting Consent Conditions : Examples from Yorkshire's Industrial Rivers

In the Yorkshire Region large volumes of water are imported from the Rivers Derwent, Ouse, Nidd, Ure and Wharfe for public supply in the industrial parts of the Region. The water is returned to the environment as effluent from sewage treatment works. Figure 3 is the residual flow diagram for the River Aire under "dry weather flow conditions", i.e., typical summer low flows - not extreme droughts. In such conditions, 50% of the river's flow in Leeds is derived from the main sewage treatment works serving Bradford. The proportion of returned water rises to 70% below the Knostrop (Leeds) STW. A similar flow pattern in dry weather is found in the catchments of the Rivers Calder, Don and Rother. Very high standards of effluent treatment are required in order to achieve the water quality objectives.

The quality of the River Aire is dominated by Marley (Keighley), Esholt (Bradford) and Knostrop (Leeds) sewage treatment works. There are few direct discharges to the river of process water from industry. Effluents from the chemicals (dyestuff, herbicides and specialist organic chemicals manufacturing), engineering, textiles, and food and drink industries go to public sewers for treatment at these works. The NWC classes are largely based on standards set for the sanitary determinands - dissolved oxygen, biochemical oxygen demand (BOD) and ammonia, which are expressed in terms of annual 95 percentiles. Figure 4 shows the 95 percentile profile of BOD and ammonia along the river, as calculated over the period 1986-89, in comparison with the EQS. The gap between the current situation and the standards must be closed and consent conditions to attain this have been calculated using mathematical water quality models, which incorporate a stoichastic simulation element for deriving frequency distributions of river and effluent flows and quality.[5,6] Figure 4 also shows the estimated profile of BOD and ammonia concentration when the water company has fully achieved the consent conditions; this is programmed for the mid 1990's.

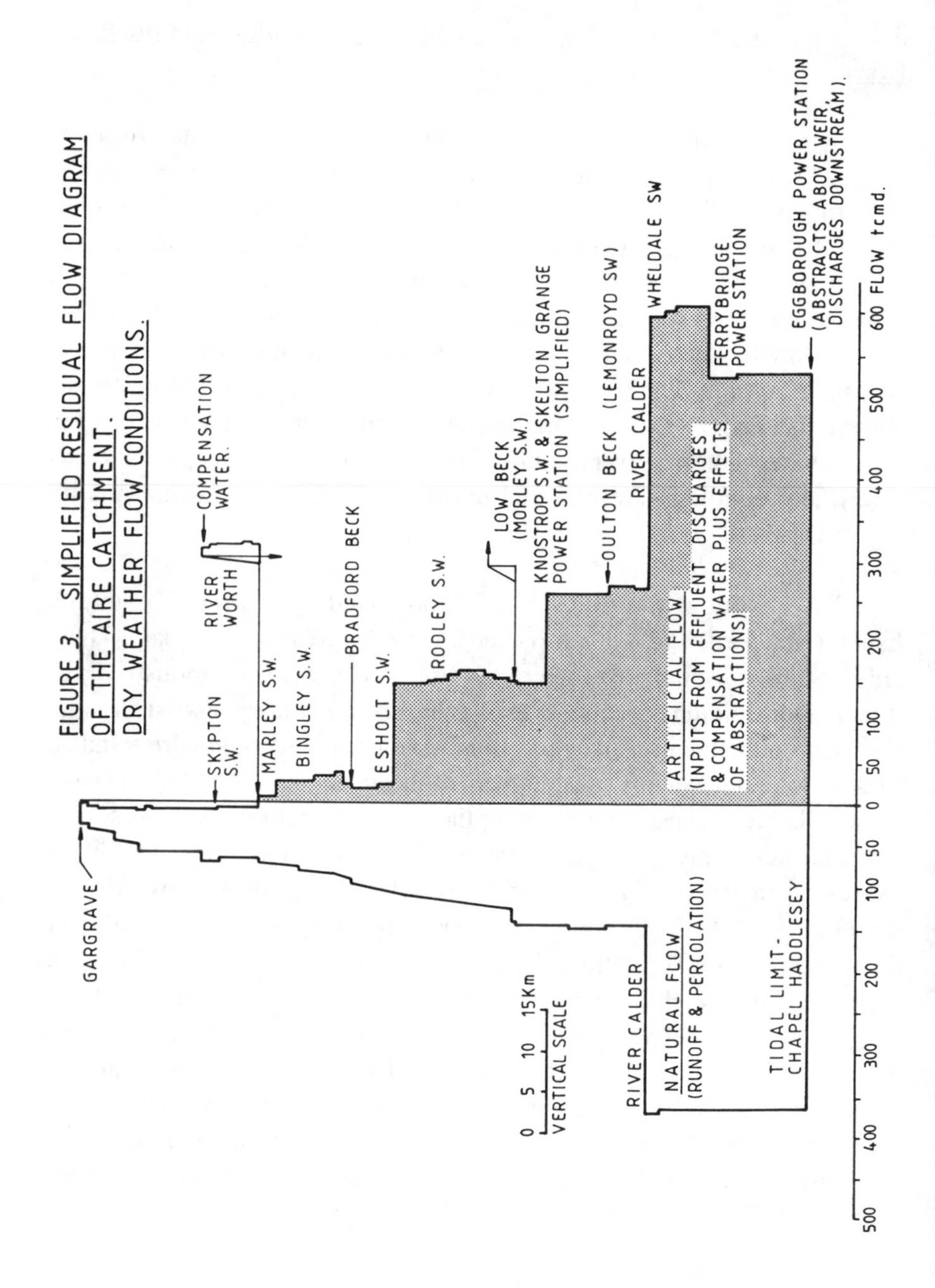

FIGURE 3. SIMPLIFIED RESIDUAL FLOW DIAGRAM OF THE AIRE CATCHMENT. DRY WEATHER FLOW CONDITIONS.

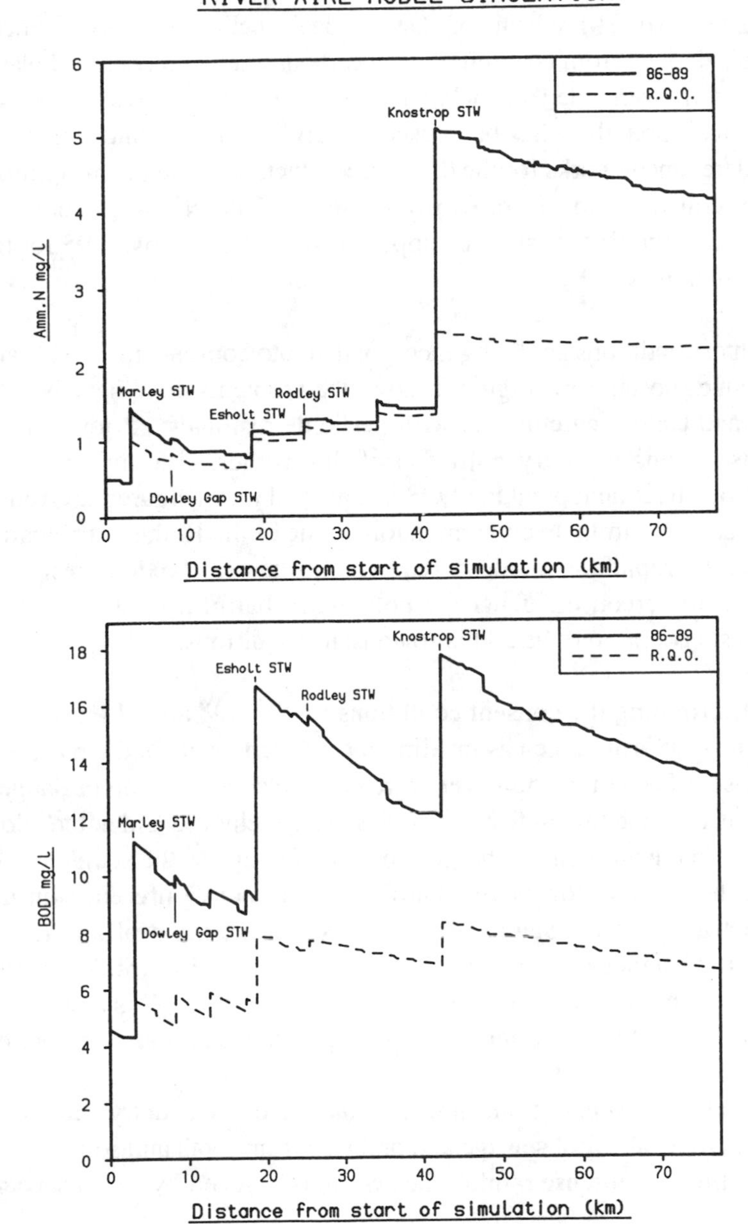
FIG.4
RIVER AIRE MODEL SIMULATION
86-89
R.Q.O.
Knostrop STW
Marley STW
Rodley STW
Esholt STW
Dowley Gap STW
Amm.N mg/L
0
1
2
3
4
5
6
0
10
20
30
40
50
60
70
Distance from start of simulation (km)
86-89
R.Q.O.
Knostrop STW
Esholt STW
Rodley STW
Marley STW
Dowley Gap STW
BOD mg/L
0
2
4
6
8
10
12
14
16
18
0
10
20
30
40
50
60
70
Distance from start of simulation (km)

The River Rother in North Derbyshire and South Yorkshire has the reputation of being one of the country's most polluted rivers with 37 km out of its total length of 51 km being of Class 4 "Bad" quality. In this catchment, pollution is derived from coal mines, coal carbonisation and chemical plants as well as from overloaded and delapidated sewage treatment works. Again mathematical modelling has been used to derive consent conditions[7] and improved treatment works for the discharges which have the greatest impact are under construction or are firmly planned. The NRA is cautiously optimistic that the Rother should support a coarse fishery by 1995 for the first time in at least 50 years.

Tighter conditions are being incorporated into consents for metals and trace organic substances. Figure 2 shows the problems there have been in the Aire and Calder catchments with pesticide residuals. Prior to 1987 there was serious mercury pollution of the Rother, although with the provision of a treatment plant the EQS is now well met (Figure 5). Action is being taken to eliminate contamination of one of the Rother's tributaries with pentachlorophenol, another "red list" substance. Toxicity testing is being used to screen effluents for potentially harmful substances and toxicity test criteria may be added to consent conditions.

In determining the consent conditions for "List I" and "List II" substances, the EQS's are taken as maximum permitted levels in the receiving watercourse. The aim is, however, to reduce the concentration of dangerous substances to the lowest that can be practically achieved and not to allow an increase in the amount discharged so as to "top up" to the standard. In addition, the Declaration of the Third International Conference on the Protection of the North Sea (1990) requires that the input of a range of dangerous substances (Appendix 2) must be reduced by 50% to 70% depending on the substance, by 1995 whether or not the EQS is met. This is the application of the "precautionary principle" to water pollution control.

Most of the sewers in Yorkshire, as in the rest of the country, are "combined", i.e., they take foul sewage and drainage from roofs and some paved areas. At times of intense rainfall the system is relieved by storm sewage

FIG.5 MERCURY: ROLLING 12 MONTHLY COMPARISON WITH THE ENVIRONMENTAL QUALITY STANDARD

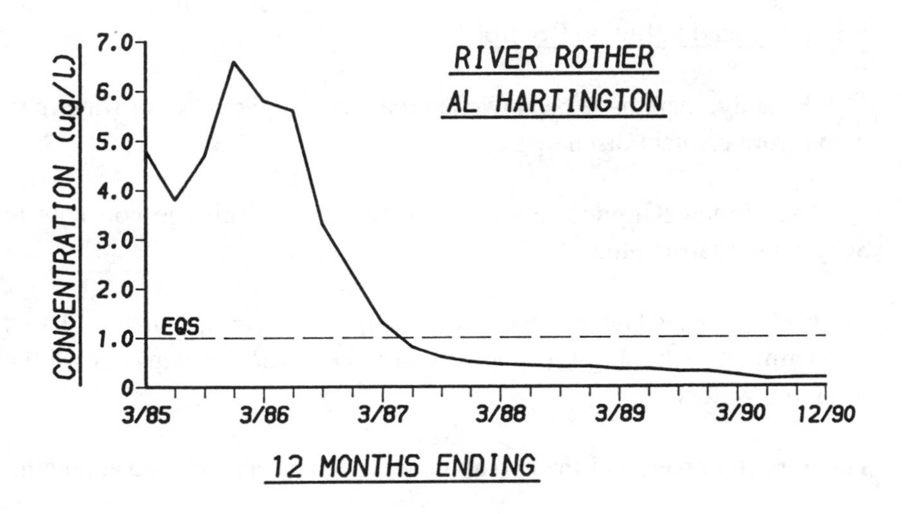

overflows. This is an acceptable practice where the overflows are designed to modern standards and correctly sited in respect to the receiving water-course. Many sewerage systems are, however, overloaded and storm sewage overflows give rise to much pollution. With most of the process effluents of industry in urban areas connected to public sewers for treatment at sewage treatment works, rather than discharged directly to river, there is concern that there could be pollution by toxic substances discharged via unsatisfactory overflows.

3.3 Integrated Pollution Control

Broadly, there have been two approaches to the control of water pollution from effluent discharges:-

a) The United Kingdom approach of tailoring discharge consents to SQO's and their related EQS's.

b) The Continental approach of setting uniform emission standards, based on treatment technology, irrespective of the nature or capacity of the receiving waters.

The respective merits of the two approaches have been debated at length.[8]

The Ministerial Declaration of the Second International Conference on the Protection of the North Sea, held in 1987, required the application of the best available treatment technologies to industrial sources of the most dangerous substances discharged to the aquatic environment. Shortly before the conference, the United Kingdom Government announced its intention to develop a new, unified approach to the most dangerous substances ("red list") which is designed to reduce their input to a minimum.

This approach requires "prescribed industrial processes" to meet emission standards based on use of the Best Available Technology Not Entailing Excessive Cost (BATNEEC) and to ensure that strict environmental quality standards are also met in the receiving waters; in effect, whichever criteria implied the more stringent controls in any particular case would apply.

This policy was brought into being by the Environmental Protection Act (1990) which introduced Integrated Pollution Control (IPC). The stated objective of IPC is to provide an approach to pollution control which considers discharges from prescribed industrial processes to all media - air, land and water.[9] The prescribed processes cover the principal sources of "red list" substances other than sewage treatment works and diffuse drainage from the land, e.g., of pesticides.[10] IPC is being introduced by the industrial sector over a four year period which commenced on 1st April 1991.

Prescribed processes will be authorised by H.M. Inspectorate of Pollution (HMIP) and these IPC authorisations will replace discharge consents. The NRA is a statutory consultee for all applications of IPC authorisation and will ensure that any necessary conditions required for water quality objectives are incorporated. A Memorandum of Understanding sets out the relationship between NRA and HMIP over water pollution control.

Trade discharges from prescribed processes to sewers also come under IPC but require also the agreement of the water utility company and will pay trade effluent charges. These controls plus tighter measures on the composition and disposal of sewage sludge, will inevitably lead in many cases to stricter standards on trade effluents discharged to sewers. Sewage treatment works themselves are not prescribed processes and the NRA will continue to regulate by discharge consents what is discharged from them and the sewerage system, into the aquatic environment.

4 POLLUTION INCIDENTS

In 1990 in the Yorkshire Region 2,700 pollution incidents were investigated, many of which were reported by members of the public. Of these, 50 were classed as being "severe" - ones which caused serious environmental damage, major fish kills or posed a risk to public water supply abstractions. The protection of public water supplies is of particular importance. For example, in 1984 the River Dee was polluted by phenol from an industrial estate further up stream. On chlorination chlorophenols were formed and over two million water customers in Cheshire and Merseyside received contaminated water with a particularly unpleasant taste.[11] Although many of the other incidents were of minor visual nuisance, e.g. an oily sheen, they indicate public concern and a demand for action.

Incidents are caused by deliberate illegal action, negligence or genuine accidents. Common types of incidents include:-

a) Treatment plant failure - caused by a breakdown, power supply interruption, poor maintenance, human error, or, in biological plants, the presence of a toxic substance which inhibits the activity of the micro-organisms;

b) Premature storm sewage overflow - badly designed overflows or overloaded sewers, sewer blockage caused by a collapse or by rubbish and rubble;

c) Pipeline rupture;

d) Spillage of oil or chemicals - tanks, drums, etc, where the bunding is absent or inadequate; over-filling of tanks by carelessness or because of a faulty contents indictor - a very common cause of pollution and an expensive waste of product as well;

e) Wash-off by rain of materials spread on land - such as sewage sludge, farm wastes, pesticides and herbicides;

f) Spray drift - from the aerial application of pesticides;

g) Storage lagoon failure - entry of tailings from mineral workings or farm slurries into watercourses, overfilling of lagoons;

h) Fires including arson - rupture of storage tanks and then wash-off by fire hoses into watercourse. The pollution of the Rhine in 1987 resulted from a fire at a factory of the Sandoz Company in Switzerland. Serious incidents resulting from fires have posed a threat to public water supply intakes on the Yorkshire Ouse. A fire at a warehouse storing herbicides and other chemicals led to a concentration of 122 000 mg l^{-1} of diquat being recorded in a stream flowing into the River Calder at Wakefield;[12]

i) Landfill operations - this can cause pollution of both surface and underground waters. It should be noted that any pollution of groundwaters is long lasting and very difficult to remedy;

j) Transport accidents - several accidents involving agricultural vehicles on the A19 north of York have given rise to a risk of contaminating public water supply intakes on the River Ouse;

k) Vandalism - an all too common cause of pollution from fires or spilt oil and chemicals;

l) Deliberate illegal disposal of wastes.

The effects of chemical pollution can be broadly summarised as follows:-

a) Aesthetic - visual nuisance caused by rubbish, solids, oil, dyes or foam;

b) Deoxygenation - removal of the water's dissolved oxygen by the oxidation of the organic matter leading to fish kills, adverse effects on other organisms and odour;

c) Toxicity - the presence of metals or organic substances which are toxic to fish or wildlife, or could contaminate water abstracted for agricultural purposes or drinking water supply for humans.

The NRA has a 24 hour, 365 day emergency service to deal with pollution incidents. The Yorkshire Region works to a target of having a Pollution Control Officer on site within one hour of the report of pollution during normal working hours and within two hours at other times. The task is to find the source of the pollution, get it stopped and where necessary, issue warnings to downstream water users. In the case of oil and some other pollutants, booms can be deployed and measures taken to contain and remove the material. Contractors are often employed on the mopping up of pollution, with the cost recharged to the firm responsible. Sometimes it is possible for the NRA to mount a fish rescue and when there is a significant fish kill, restocking is undertaken. The effects of intermittent pollution on fish and fish food organisms can be long lasting, with the benthic invertebrates taking years to recover.

The Water Act 1989 gives the NRA powers to remove or dispose of polluting matter or remedy or mitigate its presence, and to restore the water to its previous condition. Any reasonable costs incurred may be recovered. The NRA will also not hesitate to use when appropriate, its powers to prosecute those causing the pollution.

5 PREVENTION OF POLLUTION

Although the NRA has powers to control discharges, forestall and remedy pollution and to bring prosecutions, the aim must be to reduce the risk of pollution occurring in the first place. Preventative work is a very important task of the NRA. The risk of pollution is dependent on the nature of the process and the chemicals manufactured, used or stored, plus transport routes and blackspots in relation to topography, groundwater conditions and the drainage network.

Catchment control involves systematic inspection work and the development of inventories of installations which could cause a risk of pollution. Installations of known potential risk are regularly inspected by pollution control staff. Awareness compaigns are mounted and a considerable amount of education work undertaken. Adivce is given to planning authorities on pollution control measures which should be taken by developers; NRA is a statutory consultee on all applications for planning permission and licence for the disposal of waste to land. NRA is also a consultee for the site plans required by the Control of Industrial Major Accident Hazard Regulation 1984 (CIMAH) and makes recommendations on the means of reducing the risk of pollution and the protection of downstream water users in the event of a major accident.

The Water Act 1989 includes provisions which enable the Secretary of State for the Environment to introduce - by Regulations - a requirement "for those who have custody or control of poisonous, noxious or polluting matter, to take precautionary measures to prevent pollution from them". Regulations to control silage, slurry and agricultural fuel oil installations came into force in 1991 and the NRA is seeking similar Regulations to control some industrial storage facilities.

The Secretary of State can also designate water protection zones and prescribe certain activities within them, but can only do so upon application by the NRA or a Water Undertaker. A special type of protection zone is the Nitrate Sensitive Area, of which ten have been designated so far on a trial basis to protect groundwater sources in areas of intensive agriculture. No other protection zones have yet been established.

6 CONCLUSIONS

The NRA's role in water quality management includes:-

a) Advising Government on water quality objectives and standards and on other pollution control matters;

b) Regulating discharges and other sources of pollution so that WQO's are met;

c) Determining discharge consent conditions and acting as a statutory consultee on IPC authorisations;

d) Providing a pollution incident and emergency service and taking measures to mitigate pollution and restore waters;

e) Preventing pollution;

f) Maintenance of public Registers of consenting, monitoring and enforcement information.

This work is supported by a wide-ranging programme of research and development[13] and is largely financed by grant-in-aid from the Treasury and charging for discharges.

The prime instrument for managing the quality of controlled waters in England and Wales is the use-related water quality objective with its associated environmental quality standards. The discharge of waste water is controlled by consent conditions which are set in accordance with water quality standards. For prescribed processes discharge consents will be progressively replaced by IPC authorisations issued by HMIP in consultation with NRA.

Pollution incidents cause major damage to the aquatic environment and can pose a serious risk to abstractions of water for public supply, agriculture or industrial use. The NRA has powers to remedy and mitigate pollution and to prosecute offenders. The emphasis, however, must be on the prevention of pollution occurring in the first place.

Water quality management is being increasingly influenced by European Community Directives and other international obligations. New measures to strengthen pollution control are being introduced including statutory Water Quality Objectives, more rigorous specification of discharge consent conditions and the implementation of Integrated Pollution Control.

The magnitude of the task of restoring the quality of rivers which have experienced serious pollution for more than 100 years (as is the case for the Rivers Aire, Calder, Don and Rother in Yorkshire) must not be underestimated. It will take time and very large investment in sewers, sewage and industrial treatment works in the reduction of the risk and impact of pollution incidents and in the abatement of diffuse sources of pollutants including those from agriculture and contaminated land. The National Rivers Authority is however committed to improving those rivers as speedily as possible while protecting the ones which are already of high quality.

7 ACKNOWLEDGEMENT

Mr K W Newman, Yorkshire Regional General Manager, is thanked for his permission to publish this paper. The views expressed are the author's and not necessarily those of the National Rivers Authority.

8 REFERENCES

1. P.J. Newman, "Classification of Surface Water Quality: Review of Schemes used in EC Member States", Heinemann Professional Publishing, 189.
2. A.R. Agg and T.F. Zabel, *J. Water Env. Man.*, 1990, **4**(1), 44 - 50.
3. National Rivers Authority, "Discharge Consent and Compliance Policy; A Blue print for the Future", *NRA Water Quality Series*, 1990, **1**, 69.
4. National Rivers Authority, Scheme of Charges in respect of Applications and Consents for Discharges to Controlled Waters, 1991, 4.
5. K. Bowden and S.R. Brown, *Water Sci. Tech.*, 1983, **16**, 197 - 206.
6. E.A. Bramley and A.M.C. Edwards, "The Roles of Water Quality Models for Water Quality Management in Yorkshire", in "Hydraulic and Environmental Modelling of Coastal Estuarine and River Waters", R.A. Falconer, F. Godwin and R.G.S. Matthews (eds.), Gower Technical Press, 1989, 557 - 567.
7. P.W. Lai and A. Vevers, *Water Pollution Control*, 1986, **85**(3), 315 - 323.
8. N. Haigh, "EEC Environmental Policy and Britain", 2nd edition, Longman, 1989, 382.
9. Department of the Environment, "Integrated Pollution Control: A Practical Guide", 1991, 57.
10. B. Croll, *J. Inst. Water Env. Man.*, 1991, **5**(4), 389 - 398.
11. J.N. Rushbrooke and F. Beaumont, *J. Inst Water Eng. Sci.*, 1986, **40**, 173 - 192.
12. R.A. Stansfield, *J. Inst. Water Eng. Sci.*, 1982, **37**, 364 - 370.
13. National Rivers Authority, "Annual Review of R & D",1990 , 98.

APPENDIX 1

EUROPEAN ECONOMIC COMMUNITY DIRECTIVES IN FORCE DIRECTLY CONCERNED WITH THE AQUATIC ENVIRONMENT

Number	Title
73/404 & 405 82/242 & 243 86/94	Biodegradability of Detergents.
75/440	Quality of Surface Waters Intended for Abstraction for Drinking Water.
76/160	Quality of Bathing Waters.
76/464	Pollution Caused by Certain Dangerous Substances.
77/795	Exchange of Information on Quality of Surface Fresh Waters.
78/176	Waste from the Titanium Dioxide Industry.
78/659	Quality of Fresh Waters for the Support of Fish Life.
79/869	Sampling and Analysis of Surface Waters intended for Abstraction for Drinking Water.
79/923	Quality required for Shellfish Waters.
80/68	Protection of Groundwater from Pollution by Certain Dangerous Substances.
80/778	Quality of Water intended for Human Consumption.

82/176	Mercury in Discharges from the Chlor-Alkali Industry.
82/883	Surveillance of Environments affected by Waste from Titanium Dioxide Industry.
83/156	Quality of Mercury Discharges other than from the Chlor-Alkali Industry.
84/491	Quality of Hexachlorocyclohexane and Lindane Discharges.
86/280	Quality of Discharges containing certain Dangerous Substances (DDT, PCP. CTC).
87/217	Prevention of Environmental Pollution by Asbestos.
88/347	Quality of Discharges containing certain Dangerous Substances (Drins, HCB, HCBD and Chloroform).
89/428	Harmonising of Programmes for Reduction of Pollution from the Titanium Dioxide Industry.
90/415	Quality of Discharges containing certain Dangerous Substances (PER, TRI, EDC, TCB).
91/271	Urban Wastewater Treatment.
	Protection of Waters against Pollution caused by Nitrates from Agricultural Sources (agreed by EC Environment Council).
	Ecological quality of Surface Waters (Proposed).

APPENDIX 2

SUBSTANCES FOR WHICH ENVIRONMENTAL QUALITY STANDARDS HAVE BEEN SET OR ARE OTHERWISE IDENTIFIED FOR SPECIAL ACTION (FURTHER SUBSTANCES WILL BE ADDED IN DUE COURSE)

Substance	3rd North Sea Declaration Priority for Reduction	1st UK Red List	EC Dangerous Substance Directive	
			List I[a]	List II[b]
Mercury	*	*	*	
Cadmium	*	*	*	
Copper	*			*
Zinc	*			*
Lead	*			*
Arsenic	*			*
Chromium	*			*
Nickel	*			*
Drins	*	*	*	
HCH	*	*	*	
DDT	*	*	*	
Pentachlorophenol	*	*	*	
Hexachlorobenzene	*	*	*	
Hexachlorobutadiene	*	*	*	
Carbontetrachloride	*		*	
Chloroform	*		*	
Trifluralin	*	*		
Endosulfan	*	*		
Simazine	*	*		
Atrazine	*	*		
Tributyltin-compounds	*	*		*
Triphenyltin-compounds	*	*		*
Azinphos-ethyl	*			
Azinphos-methyl	*	*		
Fenitrothion	*	*		

Substance	3rd North Sea Declaration Priority for Reduction	1st UK Red List	EC Dangerous Substance Directive	
			List I[a]	List II[b]
Fenithion	*			
Malathion	*	*		
Parathion	*			
Parathion-methyl	*			
Dichlorvos	*	*		
Trichloroethylene	*		*c	
Tetrachloroethylene	*		*c	
Trichlorobenzene	*	*	*c	
1,2-Dichloroethane	*	*	*c	
Trichloroethane	*			
Dioxins	*			
PCB's	*d	*		
Iron				*
pH				*
Boron				*
Vanadium				*
PCSD's				*e
Cyfluthrin				*e
Sulcofuron				*e
Flurocofuron				*e
Permethrin				*e

Notes:
a) The "Black List" of most dangerous substances for which environmental quality standards (EQS) are set by Directives.
b) The "Grey List", EQS set by national governments.
c) EQS's come into force in 1993.
d) The Declaration has special measures for PCB's.
e) EQS's came into force in 1992.

REED BED TREATMENT OF INDUSTRIAL EFFLUENT

F.G. Holliday

ICI Chemicals & Polymers Ltd
Billingham
Cleveland, TS23 1LB
United Kingdom

1 INTRODUCTION

The "root zone" method of Professor Dr Kickuth, is a reed bed technology used in ICI's Chemical Products Area (CPA(B)) at Billingham. There is a substantial body of published material on reed bed technologies some of which was discussed at the International Conference on Constructed Wetlands held in Cambridge in September 1990. It can be said that reed beds are a natural solution to a man made problem.

First some background to the use of reed beds at the CPA(B) which was originally called Oil Works. It was opened 56 years ago by Ramsey McDonald, the Prime Minister of the day with the purpose of making aviation fuel for war planes from coal in the second world war. Over those 56 years, production has evolved into three main interests:-

a) Alcohols for plasticisers and detergent industries.

b) Phenol/acetone and derivatives for plastics, detergents, pharmaceuticals and flame retardants.

c) Amines and derivatives for drugs, detergents, paper treatment, agrochemicals and animal feedstock additives.

The CPA(B) occupies about one third of the Billingham site which is in the centre of Cleveland County and is about a mile from the sea on the River Tees. ICI built their Works at Billingham as it was situated on an anhydride deposit, was close to the Durham Coalfield for fuel and had access to the River Tees for cooling water, easy transport and sewage disposal.

Pollutants to the River Tees from the CPA(B) are now less than 0.1% with concentrations of phenol, acetone, methanol and amines decreasing in that order in more than 99.8% of the water effluent of total volume about 6 000 m^3 per day. These organic chemicals cause biological oxygen demands (BOD). Bacteria in the river feed on organic carbon chemicals, using oxygen dissolved in the water to convert the organic carbon into carbon dioxide and water, which depletes the river of its dissolved oxygen which fish in the river breath to live. The BOD load at the end of 1990 was about 6.8 tonnes per day from CPA(B). Concentrated efforts have been made since the early 1970's to reduce the BOD levels from CPA(B) in the River Tees with a reduction from about 37 tonnes per day in 1970 to about 7 tonnes per day in 1991, about a sixfold reduction.

ICI, local industry, local councils and the Northumbrian Water Authority voluntarily agreed to schemes to improve the river, which at the time was dead and smelling, so that migratory fish could once more reach the upper river. CPA(B)'s contribution to this improvement required halving the BOD load every five years, which has been achieved. Halving again over the next five years cannot be achieved easily with the methods used already: plant closures, better design, recovery and better operation as all have been fully exploited. As an example CPA(B) implemented effluent reduction schemes which cost £1.3 million from the end of 1988 to the end of 1990 to reduce BOD from 10 to 6.8 tonnes per day. What remains are undesirable organics which are purged from the products to maintain quality; so effluent treatment became important for further BOD reduction.

Studies carried out to establish off plant BOD treatment methods, although originally for the treatment of ammonia process condensate, were considered relevant to CPA(B) waste and included activated sludge and deep shaft treatment. Both suffered owing to the need for tight control of such parameters as pH, concentration, temperature etc, so that bacterial colonies were not destroyed. Re-colonising a "dead" treatment system is a time consuming operation during which untreated effluent either accumulates or is discharged untreated into the river. Comparative capital and operating costs of varying types of treatment clearly indicated that the "root zone" technique deserved further study and has since been developed into a full scale effluent treatment project.

2 ROOT ZONE EFFLUENT TREATMENT METHOD

The root zone effluent treatment method was developed by Professor Dr Kickuth from Kassel University in the 1960's. Root Zone Limited hold the exclusive rights to market his system in the United Kingdom as well as in other countries and it is with them that CPA(B) has a contract and secrecy agreement for the construction and commissioning of the reed beds at CPA(B). Professor Kickuth is the Chief Scientific Advisor of Root Zone Limited.

The system was first used in Germany to treat domestic sewage with an average input BOD of about 600 mg l^{-1} reducing to an outlet BOD of about 10 mg l^{-1}. This level of reduction certainly interested CPA(B). Up to 1985 there were some 40 systems built under Professor Kickuth's supervision to treat effluent from households, dairies, paper mills and sugar mills in Western Europe (mainly Germany). In 1990 there were 400 systems in existence.

Of particular interest to CPA(B) was the Windel Textile system, treating 1.5 km^3 per day of effluent with about 350 different organic chemical types whose BOD was reduced from about 600 mg l^{-1} to less than 40 mg l^{-1}. As a textile factory it probably used amino and phenolic chemicals, hence would have similarities to CPA(B)'s effluent.

The system can either be:-

a) A horizontal flow system with:-

Phragmites reeds planted in soil in an impervious bund, with effluent trickled in at one end and out at the other and air pumped down the reed stems by the reeds to the root hairs and out into the soil. Bacteria (the same as those in the river) live around these root hairs and carry out oxidation of carbon compounds, or

b) A vertical flow system with:-

Phragmites reeds planted as before and effluent added from pipes along the bed surface and treated effluent removed through pipes below the soil. The same bacterial mechanism takes place in both cases.

The reed has three roles, it:-

a) hosts biological colonies around the roots,
b) supplies oxygen to treat effluent,
c) develops hydraulic transport paths through the soil as the roots grow.

The soil has two roles:-

a) supplies the biological colonies; not the specific species as in most other biological treatment systems, but a wide range of different species,

b) buffers sudden effluent increases by absorption onto the soil to be released gradually with time - hence protection of the biological colonies.

3 DEVELOPMENT OF THE CPA(B) SYSTEM

Having decided to explore the "root zone" effluent treatment system as a possible means of treating the CPA(B) effluent, proof of its suitability was needed before capital sanction could be raised. The first task carried out by Root Zone Limited was to find reeds (Phragmites Australis) through greenhouse trials which could survive in the phenolic effluent. These were carried out in the Spring/Summer 1988. The effectiveness of the survivors needed to be evaluated in a pilot bed scheme where effluent was fed continuously through the trial bed over the four seasons and in different concentrations. Pilot beds were designed by Root Zone Limited, constructed in the Spring of 1988 and trials run until the Autumn of 1989. Full scale beds, designed to Professor Kickuth's methodology for the characteristics observed in the pilot beds were constructed during 1990 - October of that year saw the start of a five year commissioning programme.

The greenhouse trials showed that CPA(B) effluent killed one group of the reed sources evaluated, other groups were too affected to continue. Three groups remaining were cleared for pilot bed work. Interestingly, two of these reeds were local to ICI at Billingham, perhaps both acclimatised to the Billingham environment. The third group was that commonly used in Germany while other groups were selected by Root Zone Limited from sources around the country.

Pilot bed trials were satifactory and increased concentrations gave better effluent reductions than expected:-

0.03% phenol reduced by 90%
0.3% phenol reduced by over 94%
0.2% amines and phenol reduced by 99.1%
0.68% amines and phenol reduced by 99.9%

There was a marginal deterioration in actual discharge as was to be expected.

The full scheme was then designed and installed, costing about £5 million altogether including trials. It comprised seven 0.7 hectare beds (seven football fields) or 12.5 acres planted. Taking about 3 000 tonnes of CPA(B)'s dirty water containing about 0.15% BOD in a first phase project, which left about 3 000 tonnes containing a further 0.05% BOD to be treated in a second phase project. It was built on an old gypsum tip alongside Billingham Beck, a tributary of the River Tees.

Capital costs were high because:-

a) The tip had to be protected to avoid residues leaching out with leaking effluent. The plastic membrane for this cost about £750 000.

b) Treated effluent was returned to the River Tees rather than Billingham Beck following a desire expressed by the National Rivers Authority to clean up the beck. The pipes and pumps for this cost a further £750 000.

A considerable amount of civil engineering work was required on site as a change in direction of effluent flow was needed and although some redundant pipework could be used, drain interception was needed to direct the flow and some sources required pumping to a feed station.

Reed planting was completed last Autumn and weak effluent was fed to the reeds over the Winter. Gradual increase of both effluent strength and volume took place this year; currently full strength effluent is being fed at about one third volume. BOD reduction is about 50%.

4 TECHNICAL ASPECTS

Construction

The seven beds (100 m x 70 m with a soil depth of 0.6 m) were constructed by building clay walls on a levelled gypsum tip. Darcy's Law, modified for vertical distribution, can be applied to the construction of the beds. Total treatment area is 4.9 hectares or 12.5 acres; environmental protection being provided by a 2 mm MDPE liner with heat sealed joints.

Feed to the beds is by pump from a holding tank capable of holding half the systems effluent capacity of 3 000 m^3. This allows for buffering of any "spike" excursion. Pumps and pipework have been sized to handle the whole 6 000 m^3 CPA(B) effluent in a second phase of the project, though concentrating the effluent and removing clean water from it will allow it all to be contained in the present beds. Ten pairs of lateral distribution plastic pipes with bubblers at every 2.33 m provide effluent to each bed surface and operate on a timed sequence control along the beds to allow infiltration in a controlled manner.

Treated effluent is collected in slotted drain pipes under the soil which discharge via inspection pits into weir chambers in order that the volume of effluent in the bed can be controlled; it is then pumped to a pit for return to the site main drain.

Bed swelling, estimated at about 25 mm per year, is accommodated by bund walls which are 37.5 cm above the bed surface. Loading for this system is about 16.33 m^2 per cubic metre of effluent.

Medium

Soil or gravel/chips medium is suggested for the construction of reed beds. The soil provides the bacterial colonies which use the organic carbon effluent as feedstock whereas Gravel/chips need to be inoculated.

Kickuth's "root zone" method uses soil whose hydraulic capacity needs to be able to handle the volume of effluent requiring treatment. Hydraulic conductivity of the order of Kf 10^{-5} s^{-1} (the hydraulic conductivity constant x the flow rate) or less is developed as pore spaces collapse and/or fill with debris, but this will increase over the next three years as the reed roots punch holes/channels through the soil. Poor Kf can be compensated for by a greater bed area. Gravel/chips will typically have a Kf of about 10^{-3} hence better conductivity, but this tends to fall off fairly quickly as pore spaces fill.

Soils tend to absorb organic carbon material by weak bonding, thus shock loading is buffered in the feed part of the bed with the organic carbon being released gradually. Gravel/chips do not exhibit this effect. Buffering protects the bacterial colonies in the majority of the bed so that although shock loading will kill colonies in the feed zone, this is quickly recolonised by bacteria from the rest of the bed.

Soil of the right hydraulic conductivity was available at the ICI Billingham site for the system. Use of limestone gravel/chips has the advantage of higher hydraulic loading and being faster on line, but its biological range is reduced, it is less able to take shock loading, it has a relatively small surface area and a maintaining operation is crucial if it is not to become a polluter itself.

<u>Reeds</u>

Phragmites Australis reeds have been used in the CPA(B) system. As discussed earlier these are the survivors from the greenhouse trials. Other macrophytes are quoted in the literature as suitable to pass air to the rhizospheres, however, Phragmites Australis is the macrophyte mainly chosen by Professor Kickuth for his systems.

106 400 reeds were planted from cuttings propagated over the previous two years. Thus planting density is about two reeds per square metre though the higher the density the more air to the rhizosphere hence better treatment. Planting took place between June and October, though the literature suggests that the total planting season to be from March to October. It also suggests that the best months for planting are May and June which is borne out by the reed growth of the reeds planted earlier compared with those planted later.

Weed control has been achieved both by use of selective weedkiller and by surface flooding. Flooding requires care as new reed spikes should not be inundated or they will be starved of air. Flooding has been successful for the CPA(B) system.

It is reported that some chemicals are bound to either reeds or soil or dissipated to atmosphere from the system so work is being carried out elsewhere in ICI on how copper, vanadium and boron behave in the system.

5 GENERAL OBSERVATIONS

After almost a year's operation on the CPA(B) reed bed site, results indicate about a 50% pollutant reduction as would be expected at this stage in their development. The Kickuth "root zone" technique has been clearly proven in trial bed operation and indications are that the full scale version is following the trial bed trend.

Soil based reed bed treatment systems have both strengths:-

a) negligible sludge formation,
b) low revenue costs,
c) good secondary treatment,
d) versatility,
e) some elements become reed or soil bound

and weakness's:-

a) high land requirement,
b) long time to establish hydraulics,
c) not a polisher unless small volumes are used,
d) some elements shifted to atmosphere (but diluted dispersion by this method may be acceptable).

Root Zone Limited market the Kickuth technique in the United Kingdom, but there is a great deal of research being carried out by a number of other groups up and down the country. However, Root Zone Limited designed and constructed the "root zone" systems for ICI CPA(B) under a turn-key contract arrangement. Projection of the BOD reduction with reed bed technology shows a significant drop to below one tonne per day when all the effluent is being treated from CPA(B).

6 CONCLUSION

Reed beds are ecologically very attractive as they attract a lot of wild life. The sludge contained within the beds attracts a variety of detritus eaters (worms, slugs etc) which, in turn, together with the reed seed attracts birds and small animals. A fox has even been seen foraging on the trial reed beds.

Running costs of reed beds are low, allied more to agricultural than industrial operation and they are robust, being able to tolerate wide variations in feed stock composition without permanent damage or disruption to treatment. However, they do require a large land acreage, have a long lead time and are hydraulically limited.

By the year 2000 the CPA(B) will have no effluent as it will all be treated on full scale reed beds as the quality of water from the present reed beds has shown the viability of the process. The clean water produced will be re-used in cooling water systems

IN BALANCE WITH NATURE

G.L. Angell

BASF plc
Hadleigh
Ipswich
Suffolk, IP7 6BQ
United Kingdom

In the beginning there was earth, beautiful and wild: and then man came to dwell. At first, he lived like other animals feeding himself and his family on creatures and plants around him. And this was called in Balance with Nature.

Soon man multiplied. He grew tired of ceaseless hunting for food: he built homes and villages. Wild plants and animals were domesticated. Some men became farmers so that others might become industrialists, artists or doctors and so on. And this was called Society.

Man and society progressed. With his God-given ingenuity, man learned to feed, clothe, protect and transport himself more efficiently so he might enjoy life. He built cars, houses on top of each other, and he produced nylon. And life became more enjoyable. The men called farmers became efficient. A single farmer grew food for the families of 28 industrialists, artists and doctors, and writers, engineers and teachers as well.

To protect his crops and animals, the farmer used substances to repel or destroy insects, diseases and weeds. These were called pesticides. Similar substances were made by doctors to protect humans. These were called medicines. The age of science had arrived cleaving society in two.

The larger part lacked the skills and resources needed to benefit from science and its ever multiplying families continued to live in squalid conditions widely exposed to famine and disease. The smaller part harnessed science and technology and thus enjoyed a better diet and longer and happier lives yet within its ranks there were increasing numbers who disapproved of the farmer using science. They spoke harshly of his techniques for feeding, protecting and preserving plants and animals. They castigated him for upsetting the balance of nature and they yearned for the good old days. This had strong emotional appeal. In the meantime farmers had become so efficient that society gave them a new title - unimportant minority.

So it came to pass that this more prosperous and developed arm of society which had neither recent experience nor any conception of food shortages, passed laws abolishing pesticides, fertilisers and food preservatives. Insects, diseases and weeds flourished, crops failed and animals died.

In order to survive industrialists, artists, doctors and the rest were compelled to grow their own food. They were not very efficient. Agricultural land became so unproductive that people and nations fought each other to gain more of it. Millions perished. The few survivors lived like animals feeding themselves on creatures and plants around them. This, as in the beginning of the story, was called "in balance with nature".

Far fetched? Not necessarily, although there may be a more realistic scenario; sufficient if the story illustrates its central point.

No one would disagree with the objective of protecting and improving the environment, to ensure that food production methods provide wholesome and safe foods and that irresponsible use of non-renewable resources are avoided. But in all these situations there are costs and benefits which need to be evaluated to arrive at the most environmentally appropriate decision.

It is now two generations since the public faced the spectre of food shortages, so today there is already complacency and a lack of awareness of the very considerable difficulties faced by farmers. In addition, the public is continually bombarded by the media about the dangers from food infections, such as salmonella and listeria and problems of intensive food production, including the use and application of chemicals. In this respect the most depressing predictions are made about the future unless there is a "return to Nature!".

What exactly is meant by this and how it could be brought about is never clearly explained. Terms such as "chemical free food" are bandied about without any understanding of what is implied. Although such quotations may sound quite ridiculous to some, they are highly emotive and, in turn, closely linked to the emotional question of health. The sheer column inches and television viewing time devoted to diet, food infections, intensive farming practices and nitrates in water demonstrate how widely food production is used as a battleground - and not only for the hearts and minds of those who are genuinely concerned. Public pre-occupation with food scares encourages anarchists and blackmailers to sabotage leading brands of food and drive them off supermarket shelves. It also helps to target the campaigns of activists against intensive systems.

The regular purveyors of dietary doomwatch rely upon prediction as their main weapon. It is so difficult to challenge, simply because one cannot prove a negative. A prediction allows tremendous licence because one is unlikely to be called to account for even the most extreme statements, and the more extreme these are, the more likely they are to catch the headlines. Rebuttals will immediately attract the Mandy Rice Davis syndrome (i.e.) "they would say that wouldn't they".

Nevertheless, let us look at how these statements stand up. Let us begin with the word "natural". What is natural? Firstly, the word natural is not synonymous with wholesome and safe. In fact, natural toxins produced in growing crops are far more common than man-made chemicals and on average humans ingest 10,000 times more natural toxins than man-made pesticides.

What is a toxin? A more common word is "poison". Everything is a poison. Nothing is a poison. It is the dose that makes a poison. But all things are relative! For example, it is much better to go into a supermarket and buy food produced within the maximum residue limits than find a cockroach or caterpillar that may be carrying 40 or 50 disease organisms on its feet.

Perhaps because it gives people the perception of an idyllic landscape, a great deal is heard about so called "organic production" being environmentally friendly, and yet when one considers that up to twice as much can be produced intensively from half the amount of land, it is difficult to understand how this can be so. Just one example; would the nitrate leakage from 100 acres of organically farmed land be less than from 25 acres farmed intensively? Surely it would be more sensible to farm the better land more intensively and devote the poorer land to conservation!

Alternative methods of pest and disease control although attractive should also be approached with care; for example, breeding plants with natural disease resistance depends on producing toxins within the plant, i.e. transferring the factory into the plant. There is no logical reason why the "natural" toxin approach should be safer than the synthetic approach, particularly when one considers the huge safety assessment programmes to which pesticides are subjected. The toxins on which natural pests depend would not normally be tested at all for safety.

What is the position with man-made chemicals in food production and storage? Research and development is aimed at making them less persistent in the environment and more effective in smaller quantities. If a new candidate shows potential environmental problems, it is dropped from development. In addition they are applied with greater knowledge of and concern for the risks. They are also much more strictly regulated by national Governments and international organisations; for example, in the United Kingdom allowable levels of residues in food are set with a hundred-fold safety margin and tolerances so low that few pesticide residues leave the farm.

As present forecasts suggest that by the year 2025 the world population will double, all the help that can be raised will be needed to be able to produce sufficient food. These problems, as with so many of the difficulties that humanity faces on this planet, can only be solved by science and technology if the "horn of plenty" is to be retained and extended to other areas where famine already exists.

Ironically, this choice of plenty which is enjoyed in the developed world is at the root of the problem simply because man who has food has many problems, whereas man who has no food has but one! But, in many other areas society seems to accept that mankind should continue to advance. It is not suggested that because people are living longer, more and better medicines should not be produced.

How can farming help the public appreciate their present food supply? The results which science and technology have so far achieved should be used to explain that people are healthier and better fed than has ever been and that the average life span has increased within the last hundred years from under 50 to over 70. Alongside these facts other facts can be set about what could happen to mankind without the aid of such technological tools. To this, thanks to Perestroika, can be added reliable data on what can happen to food production if the fruits of science and technology are stifled or squandered in the toils of bureaucracy. This factual analysis should enable the public to evaluate more clearly the advantages of today, and hopefully count their blessings.

So farming in the United Kingdom can be, and should be, rightly proud of its achievements rather than apologise for them. Of course the public's voice should be heard, but equally the public should consider the facts and try to separate logic from emotion, otherwise the real food and farming issues will not get a proper hearing.

IT'S OUR DUTY TO CARE

P.J. Howard

Leigh Environmental Limited
Four Ashes
Wolverhampton
Staffordshire, WV10 7BQ
United Kingdom

The Environmental Protection Act 1990 received the Royal Assent on the 3rd of November 1990. It is basically an enabling act introducing a series of regulations. Part Two of the Act concerns the issue of waste on land and affects the actions of waste producers and their relationship with the waste management industry. One of its objectives is to improve the standards of contractors operating in the industry. Some of the changes include the ending of the hitherto dual role of operator and regulator for local authorities via the introduction of local authority waste disposal companies (LAWDCs).

The licencing of waste management operations must now give due consideration to whether the applicants are "fit and proper" in the terms of their past conduct, i.e. prosecution record, technical competence and financial viability. This is to prevent instances when operators can abandon a site and leave the local authority or national government with the problem of cleaning up and restoration. This is further strengthened by legislation preventing surrender of the site licences without the approval of the waste regulation authority. Further, it creates new duties on waste regulation authorities to monitor sites where no licence is in force for harm or risk associated with waste deposits.

Another major element of the Environmental Protection Act is the concept of "duty of care". This concept was first mentioned in 1985 in the Royal Commission on Environmental Pollution (Eleventh Report) entitled "Waste Management - The Duty of Care". Section 34 of the Act creates a duty of care with respect to waste. The duty applies to anyone who imports, produces, carries, keeps, treats or disposes of controlled waste or who has control of such material as a broker. These people are under a duty to take all reasonable measures applicable to them to prevent others from treating waste materials in an unauthorised or harmful manner. They must prevent the escape of waste from their control and take responsible steps to ensure that it does not escape from the control of others. It is also a duty to ensure that the waste is transferred to an authorised person. Further regulations relating to the registration of carriers will re-inforce the duty. Finally, the waste must have a sufficient written description attached to it when transferred from person to person during its disposal route. Breach of these regulations will constitute a criminal offence and may also give rise to civil liability if damage can be proven to have resulted.

Thus detailed attention will need to be concentrated on the nature of the waste materials produced, their secure site storage whilst awaiting disposal and the contractual arrangements with a duly registered waste disposal contractor.

The waste management industry is accepting the new challenge by providing advice and analytical expertise to waste producers in determining a safe disposal route and information required to complete its transfer from source to final disposal point. Responsible waste disposal contractors operate laboratory facilities that enable the composition of the waste material to be determined accurately, whether it be the concentration of metals using ICP or the fingerprinting of organic compounds using GC-MS techniques. These central facilities will be backed up by site laboratories which have the capability of determining that the waste material delivered to the site conforms to the initial analysis. They will also provide analytical functions necessary for monitoring the performances of the site operations.

Before considering off-site disposal, a waste producer must always examine ways of recovering, recycling or reducing the waste material. In this context, the installation of in-house treatment plants needs to be examined. Such plants can be operated by the waste producer or in some instances waste management contractors. Recovery and re-use is also given first consideration by waste management companies.

Oil/water emulsions are usually separated using various techniques, which may include "chemical cracking", ultrafiltration and vacuum distillation to produce a base oil which can be re-refined or blended to create a useable product. Solvents represent another class of waste materials that are recovered and re-used.

Chemical methods for waste treatment and disposal are many, varied and use several different principles. These include:

a) neutralisation - whereby acidic effluents are neutralised to precipitate various metal hydroxides from solution. These can then be coagulated and separated from the effluent to produce a filter cake acceptable to landfill and liquors suitable for discharge to industrial sewers.

b) incineration - whereby organic effluents are incinerated at high temperatures. The techniques employed in this option are highly sophisticated, particularly in the control of emissions.

c) stabilisation - whereby waste materials are mixed with cementatious materials to render them suitable for discharge to landfill sites.

e) encapsulation - whereby suitable organic waste materials are mixed with epoxy resins to give an organic polymer based solidification product.

f) vitrification - whereby waste materials are fused with silicates to create a stable product. This technique has found extensive use in the nuclear industry.

g) oxidation - whereby organically contaminated aqueous wastes are reacted with oxygen at elevated temperatures and pressures to produce a liquor which is suitable for biodegradation.

Newer technologies under consideration include ion exchange and electrochemical techniques, e.g. the so called "silver bullet" technology and absorption techniques involving the use of activated carbons. Many of these technologies have found their origin in the treatment of nuclear waste and as such have always been very expensive to operate. However, changes of attitude promoted by the introduction of the Environment Protection Act is making them worthy of consideration.

The waste management industry is ever developing techniques for the handling and disposing of controlled wastes in a safe and reasonable manner and as such will continue to work in close co-operation with waste producers to ensure that the principles in the Environment Protection Act are met.

ENVIRONMENTAL IMPACT OF THE UNITED KINGDOM NUCLEAR FUEL REPROCESSING INDUSTRY

D. Jackson

British Nuclear Fuels plc
Sellafield
Seascale
Cumbria, CA20 1PG
United Kingdom

1 INTRODUCTION

Within the United Kingdom (UK), nuclear power currently generates some 64 000 GWh per year of electricity, or 21% of the total national electrical capacity. World-wide, some thirteen countries produce between 25 and 75% of their electricity from nuclear power. Overall, nuclear power generates electricity equivalent to that from 1 000 million tonnes of coal per annum. This is more than the total world electrical generation in 1956, when the first commercial sized nuclear power plant opened at Calder Hall, on the Sellafield site in Cumbria. By the end of this century plant already on order will bring world-wide nuclear power generation up to 1 350 million tonnes coal equivalent per year. The tight regulation of the industry ensures this has been achieved with minimum intrusion into the environment and without adding to acid rain, nitrogen oxide emissions or the greenhouse effect.

Identified, commercially extractable, world resources of uranium guarantee a long-term future for reactors whilst reducing reliance on fossil fuels, which have many other uses. Reprocessing spent nuclear fuel allows recovery of uranium remaining in the fuel bar at the end of its useful life, up to 96% of the original, for subsquent re-use. This reduces dramatically the total amount of waste arisings.

Discharges of those wastes which do arise during the operation of power stations, or during the fuel manufacture and reprocessing cycle, are permitted only in accordance with Certificates of Authorisations. These are granted through the Radioactive Substances Act (1960) by the Department of the Environment (DoE) and the Ministry of Agriculture, Fisheries and Food (MAFF), for sites in England and Wales, and by the Scottish Office (SO) for sites in Scotland.

In establishing discharge limits the Authorising Departments take into account the radioactive waste management objectives of the Government. These are currently:[1]

a) all practices giving rise to radioactive wastes must be justified, i.e. the need for the practice must be established in terms of its overall benefit;

b) radiation exposure of individuals and the collective dose to the population arising from radioactive wastes shall be reduced to levels which are as low as reasonably achievable (ALARA), economic and social factors being taken into account;

c) the effective dose from all sources, excluding natural background radiation and medical procedures, to representative members of a critical group should not exceed 1 mSv in any one year. However, effective doses up to 5 mSv are permissible in some years provided that the total dose does not exceed 70 mSv over a lifetime.

In addition, the Radioactive Waste Management Advisory Committee (RWMAC)[2] and the National Radiological Protection Board (NRPB)[3] have advised that, for all nuclear site discharges, an objective of waste management practice should be that the committed effective dose to the critical group, related to each year of operation, should be no greater than 0.5 mSv. The Government has accepted this advice.

In order to ensure that these objectives are met, there is a statutory obligation under the terms of the discharge Authorisations to carry out defined monitoring programmes both for discharges and for environmental radioactivity. Independent monitoring programmes and dose measurements are also carried out and reported by the Authorising Departments[4,5,6,7,8] and other groups.[9,10]

2 MONITORING OF THE ENVIRONMENT

Within the UK, most spent nuclear fuel is sent to the British Nuclear Fuel plc (BNFL) reprocessing facility at Sellafield, Cumbria. As a consequence, the major part of the industry's waste arisings are concentrated at this site, which thus forms the focus for assessing the environmental impact of the UK nuclear fuel operations. All high-level and intermediate level wastes are currently stored on this site. Potential losses to the environment are reduced by conversion to vitrified or encapsulated products respectively. These account for more than 99% of all wastes separated at Sellafield. The remaining low level waste is discharged to the environment in one of three forms: liquid, airborne or solid.

Over the past fifteen years, considerable resources have been devoted to reducing the activity contained in liquid and airborne low-level discharges. Currently the Enhanced Actinide Removal Plant (EARP), costing £240 million, is being commissioned to reduce discharges further; primarily of plutonium and americium.

In order to assess the impact and acceptability of these discharges, BNFL conducts extensive environmental monitoring around its Sellafield and Drigg sites (Figure 1). Several thousand samples are taken annually at a cost of £3 to £4 million including analysis. Results indicate, broadly,

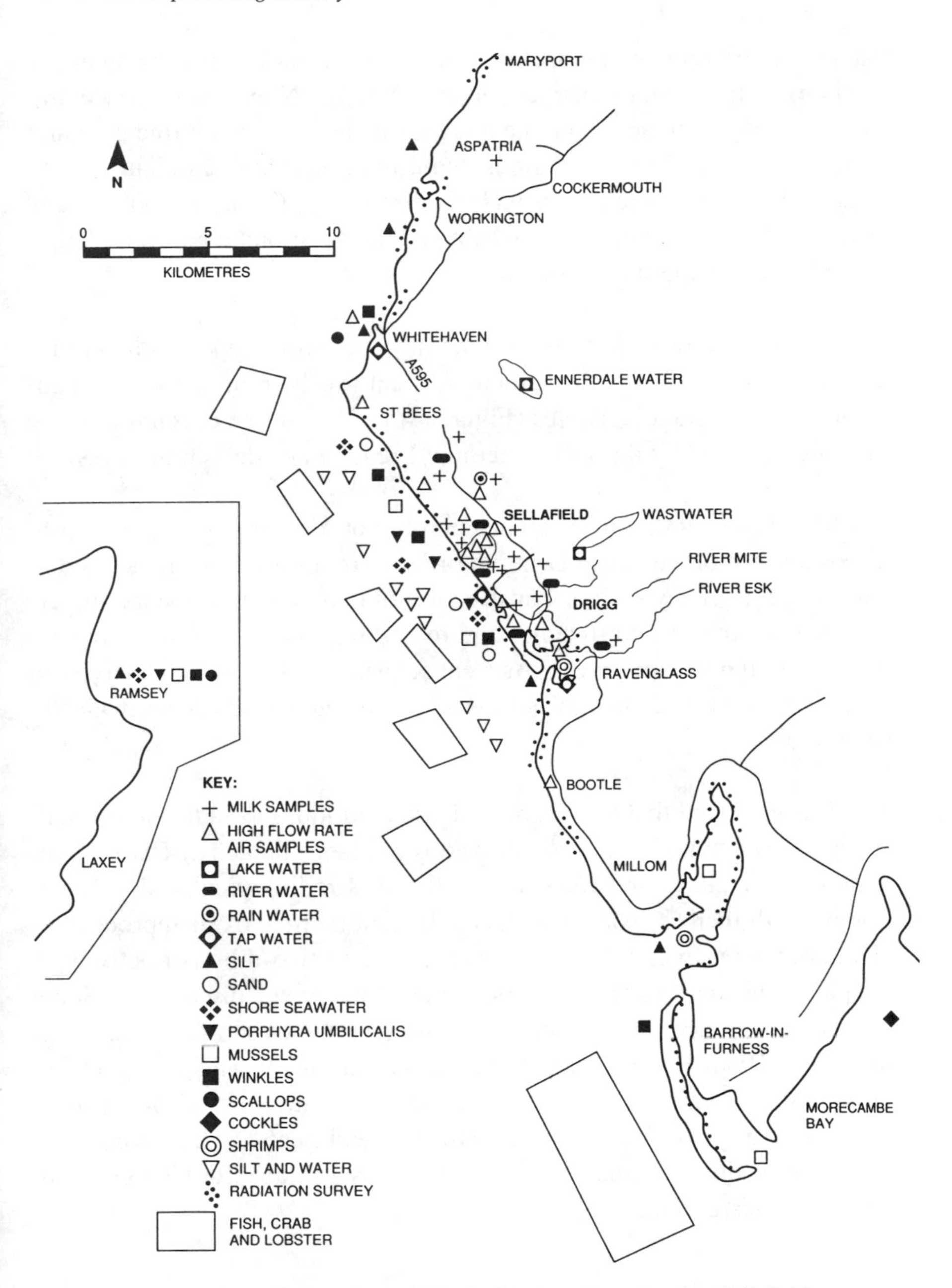

Figure 1 Environmental Monitoring Locations local to Sellafield

that concentrations of radioactivity in the local marine environment are declining in line with discharges (Figures 2 & 3). Nonetheless, concentrations of many radioactive elements remain elevated in the fine silts and sediments of the intertidal regions, leading to measurable radiation dose rates. These levels are also reducing more slowly (Table 1) and in many areas are now almost indistinguishable from background; generally 0.05 - 0.15 μSv h^{-1} for these locations.

Within the terrestrial environment, concentrations of radioactivity have always been low. Nonetheless, local levels in milk and air again broadly reflect site discharges (Figures 4 & 5) with the exception of the influence exerted by the 1986 Chernobyl reactor accident in the USSR.

BNFL has further invested in a number of measures at its solid waste disposal and storage site at Drigg in order to minimise intrusions into the environment. They include construction of concrete lined vaults, impermeable caps over filled trenches and re-routing of drainage waters away from the Drigg stream to sea. As a consequence, levels of radioactivity in the Drigg stream and other local waters have recently declined dramatically (Figure 6).

It is expected that environmental concentrations of radioactivity will continue to decline slowly over the next few years although discharges will remain more static and for some of the less radiologically significant isotopes will increase somewhat when the new thermal oxide reprocessing plant comes on-line in 1993. At present up to two-thirds of activity in samples of biota arises from historic, rather than current, discharges. Some of this activity decays at a rate that is appreciable over a few years (e.g. ruthenium-106 half life 1 year; strontium-90 half life 29 years). Dispersion throughout the environment acts to reduce concentrations of longer lived isotopes (e.g. plutonium and americium in sediments), whilst some elements are gradually sequestered to become less available for biological uptake (e.g. caesium binding in soil).

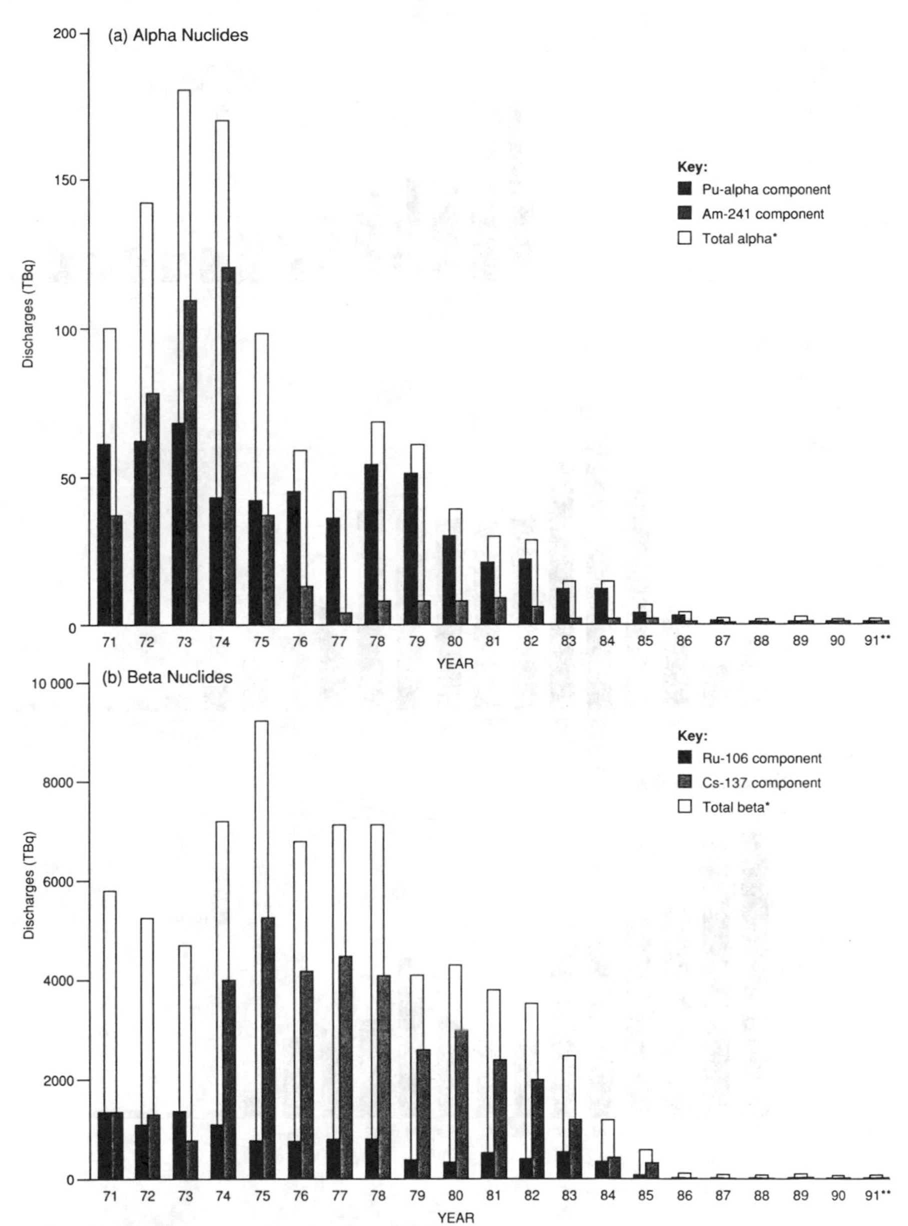

* Total alpha and total beta are overall control measures which do not reproduce precisely the contribution of individual nuclides.

** Estimated only

Figure 2 Principal Liquid Discharges from Sellafield

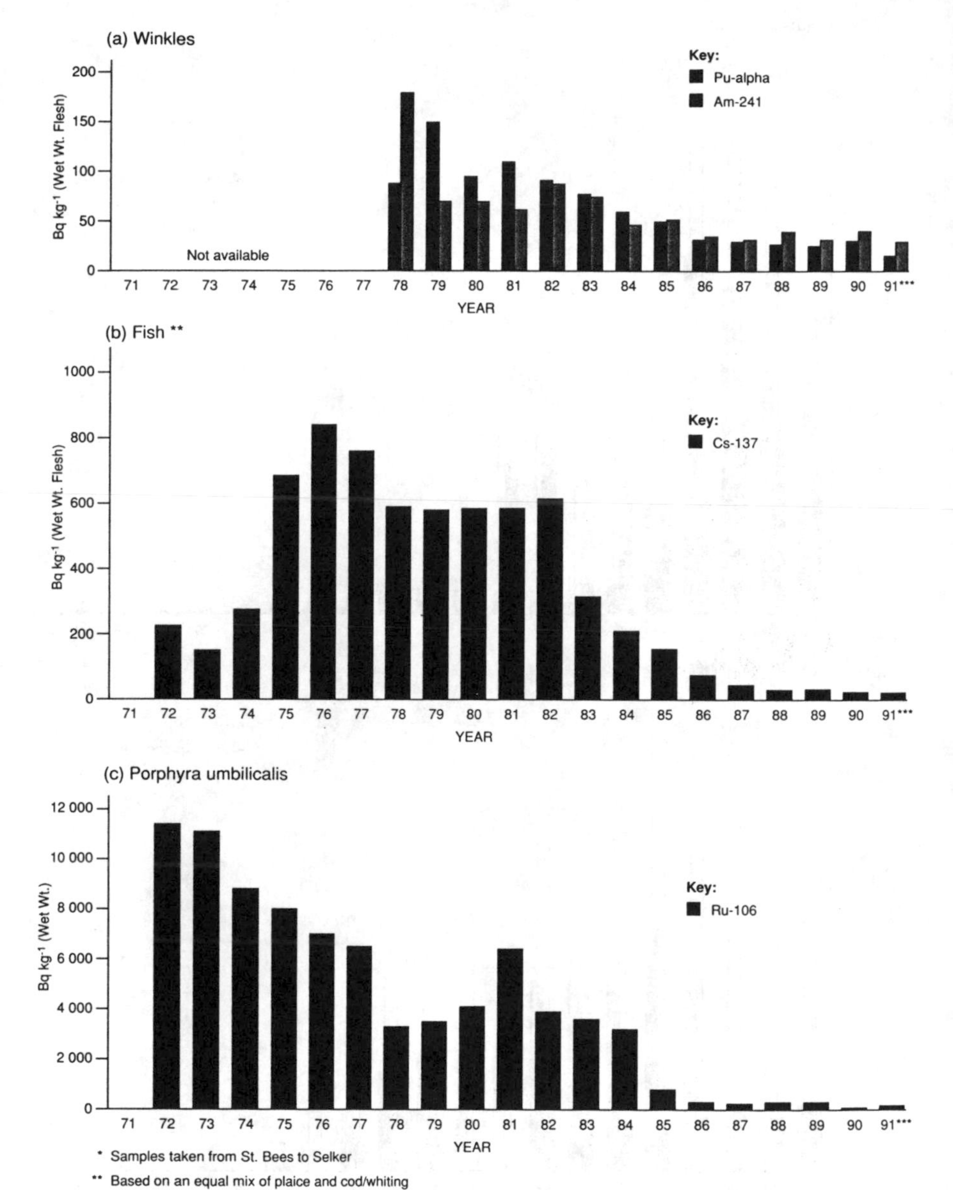

* Samples taken from St. Bees to Selker

** Based on an equal mix of plaice and cod/whiting

*** Results are for January to March only

Figure 3 Radioactivity in Marine Biota near to Sellafield*

Table 1 Radioactive dose rates over intertidal regions in the vicinity of Sellafield.

Location	Substrate	Dose Rate/μSv h^{-1}						
		1972	1976	1979	1981	1984	1988	1991*
Whitehaven Harbour	silt	-	-	-	0.6	0.7	0.5	0.24
Raven Villa	salt marsh	2.4	2.0	1.3	0.7	0.5	0.4	0.3
Eskmeals Viaduct	salt marsh	2.8	2.5	2.0	1.1	1.0	0.7	0.32
Seascale	sand	-	0.4	0.45	0.22	0.2	0.16	0.15
Sellafield	sand	-	0.35	0.35	0.35	0.21	0.2	0.14
St Bees	rock	-	0.6	0.5	0.5	0.2	0.19	0.14

* - average for January to March only

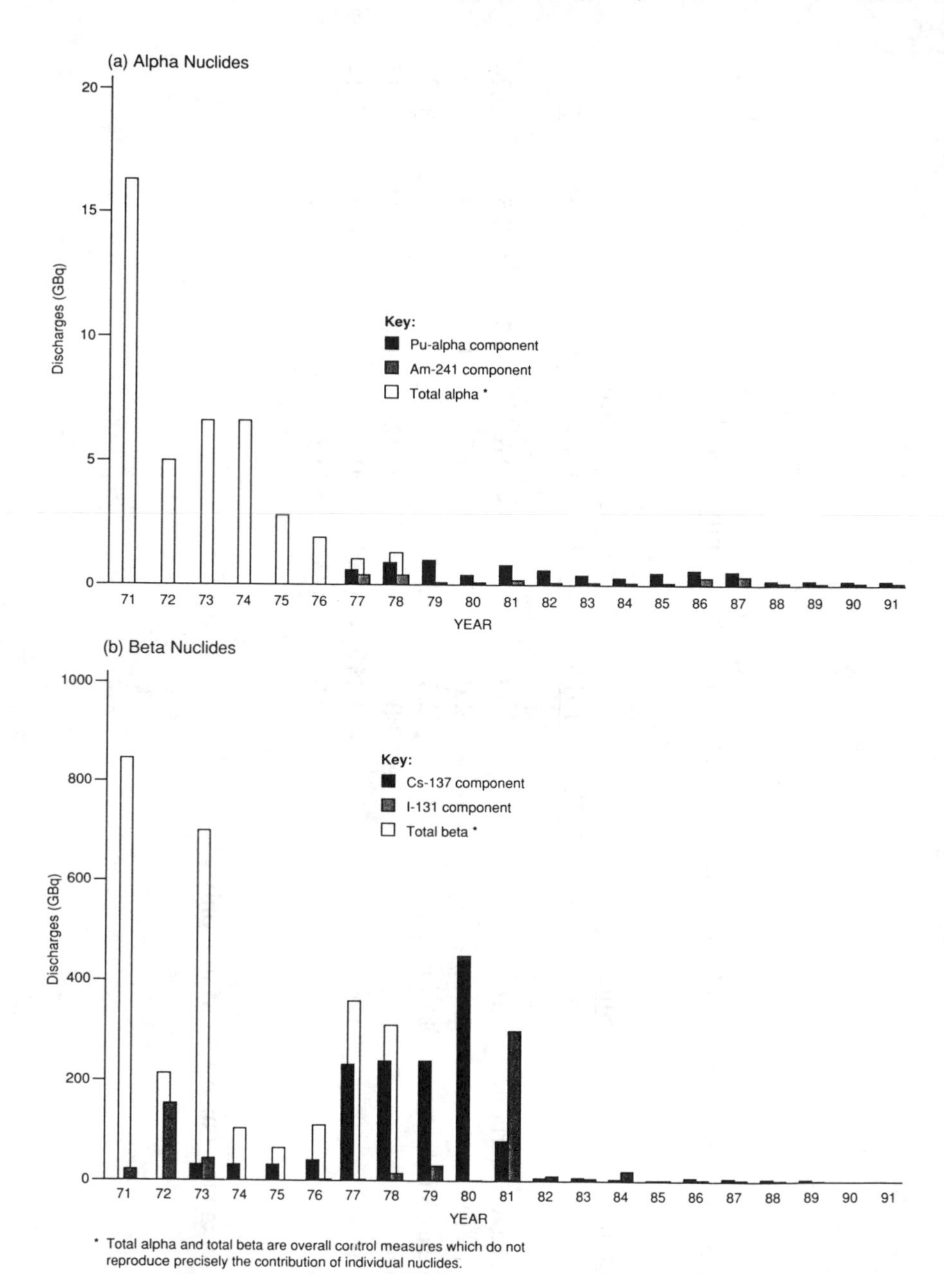

* Total alpha and total beta are overall control measures which do not reproduce precisely the contribution of individual nuclides.

Figure 4 Principal Discharges of Radioactivity to Atmosphere from Sellafield

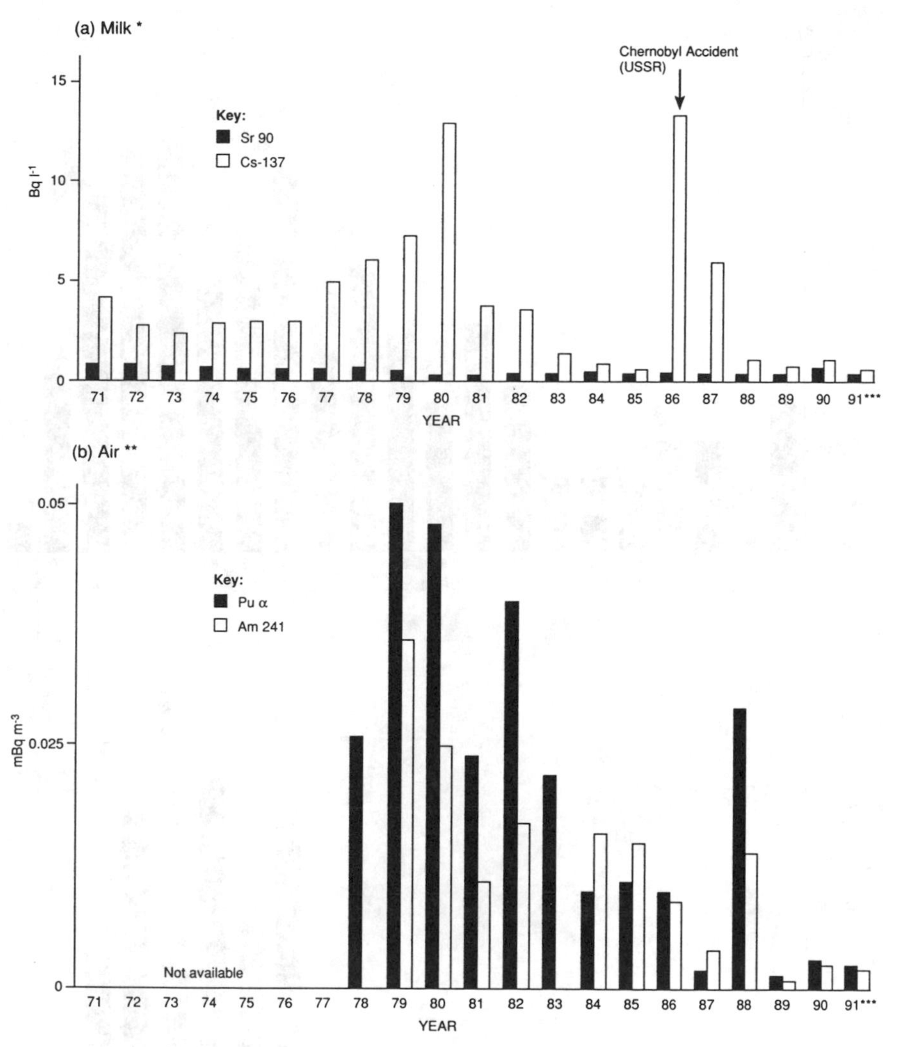

* Farms 0-3km from Sellafield

** Air samplers at the site perimeter

*** Results are for January to March only

Figure 5 Concentrations of Radioactivity in Air and Milk near to Sellafield

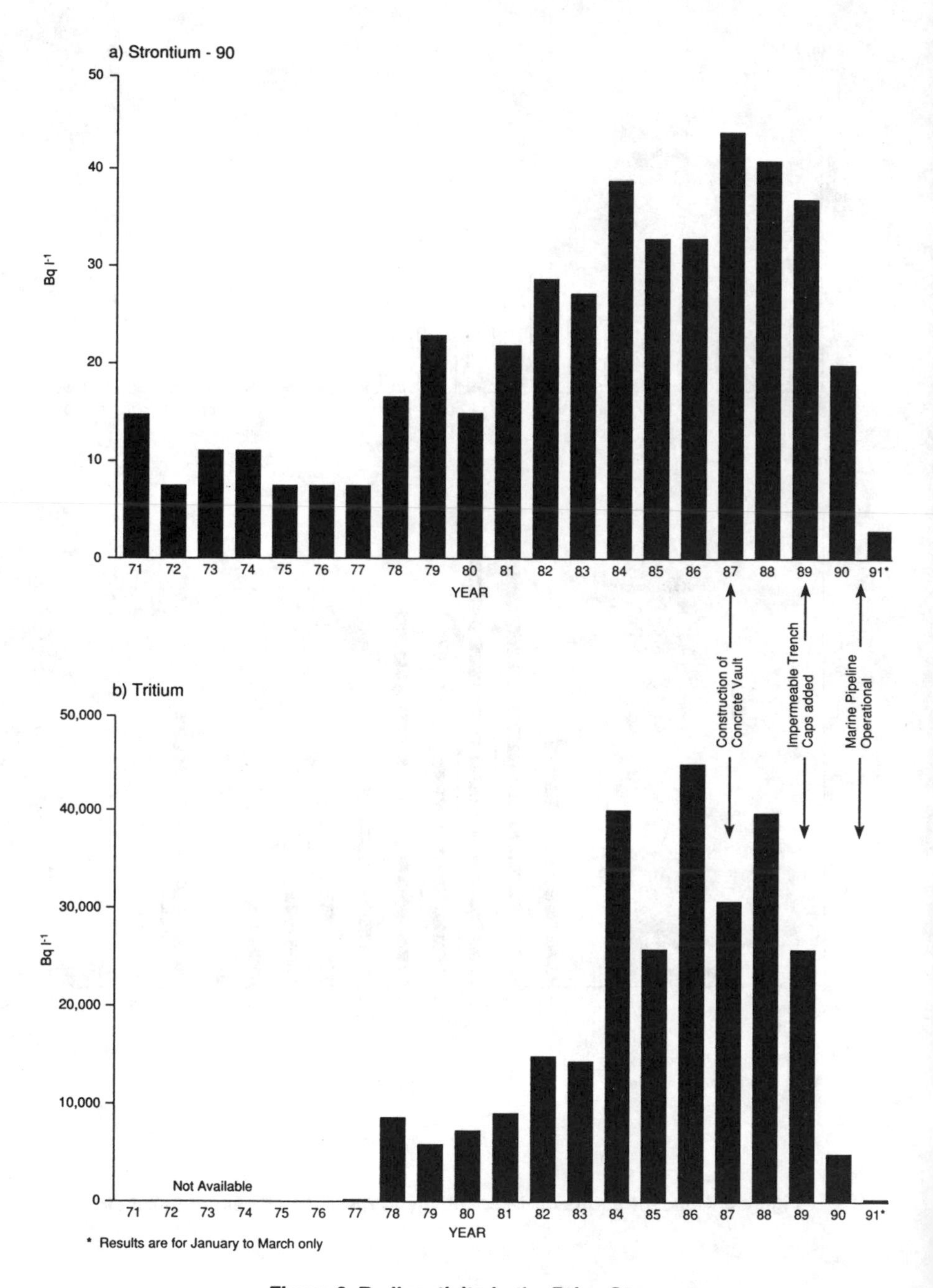

Figure 6 Radioactivity in the Drigg Stream

3 IMPACT ANALYSIS

A key concept for assessment of impact to the public is the "critical group".[11] This represents a small number of people with identified habits, such as consumption of specific foods at a high rate and/or occupancy of certain areas, constituting "critical pathways", which result in them being the most highly exposed members of the public. From this, it follows that where radiation doses to members of the public are being assessed against limits or targets, the critical group dose represents the most stringent point of comparison.

Data identifying critical groups are provided by MAFF and SO for sites in England and Wales and in Scotland respectively based on survey work[6,7,8,12] and supportive studies undertaken by operating sites.[13] A number of such groups may be identified, for particular forms of waste discharge (e.g. to sea or atmosphere) or related to specific types of radionuclide (e.g. where external or internal radiation exposure may dominate). The radiation doses to each critical group are assessed using relevant measurements of radioactivity from the monitoring programme and published data on the metabolic behaviour of radionuclides.[14,15,16]

In some instances individuals who are members of one critical group may also feature in another such group. It is thus necessary on some occasions to add the doses received by two or more identified groups.

In addition to estimating doses to critical groups, it is necessary to estimate radiation doses to populations as a whole if a comprehensive picture of environmental impact is to be obtained.[17] This involves the concept of "collective dose"; that is the summation of all the individual radiation doses received by a population over some defined period of time. Since radionuclides persist in the environment, subject to processes of dilution, dispersion and decay, the public will continue to receive radiation doses (albeit at a decreasing rate) for some time after a discharge is made. Calculating the collective dose therefore involves tracing the behaviour of radionuclides over extended periods following the discharge.

In practice, collective doses are dominated by the summation of a large number of exceedingly small doses received by individuals who are remote, in both space and time, from the point of discharge. In consequence, the calculation of collective dose rests largely on the use of theoretical models which predict the dispersion of radionuclides over large geographical areas and timescales. The unit for collective dose is the "man-Sievert", which emphasises that the value quoted is the sum of doses received by a number of individuals.

4 CRITICAL GROUP DOSES

Critical groups have been identified for marine, atmospheric and terrestial pathways, relating to consumption and occupancy habits. In addition, a number of hypothetical critical groups are derived, based on potential pathways, historical habits or additivity of existing sub-groups. Dose estimates depend on a number of factors including concentrations of radionuclides in the environment, quantities ingested or inhaled or time spent in an area and transfer of radioactivity. The main pathways identified relevant to discharges from Sellafield are:

a) high rate consumption of seafoods (especially fish and shellfish) and of agricultural produce (especially milk);

b) external radiation from exposed intertidal sediments or from deposition to ground;

c) inhalation of and exposure to airborne radioactivity.

In general, concentrations of radioactivity in the marine environment reflect liquid discharges through the sea pipeline, whereas the terrestial environment reflects atmospheric discharges. Some crossover does occur with, for instance, sea to land transfer processes, but this does not have a major effect.

The marine foodstuff critical group consists of a number of people in the Cumbrian coastal community consuming fish and/or shellfish caught in the Sellafield area between St. Bees and Selker. Consumption rates appropriate to critical groups and other local consumers are presented in Table 2. For assessment purposes it is assumed that a single group of people consume fish, molluscs and crustaceans at these rates. Although the above average consumers of one sub-group are not generally members of another sub-group, this procedure is not excessively conservative and critical group doses are estimated currently to be in the range 125 - 175 μSv. No allowance is made in this estimate for food preparation losses of radioactivity, such as the immersion of winkles in water overnight for gut clearance. This practice is recommended to reduce the hazards of contamination by sewage and industrial effluents and would reduce the dose estimate by a factor of approximately two.[18]

Table 2 Assumed consumption rates for critical group members near to Sellafield.

Foodstuff	Consumption rate*/kg per year
milk	300
eggs	30
beef	60
lamb	30
offal	20
poultry	30
root crops	120
green vegetables	80
fruit	60
fish (cod & plaice)	36.5
crustaceans (crab & lobster)	6
molluscs (winkles)	8.3

* - consumption rates estimated for adults.[3,7]

This group may also receive doses from other pathways such as inhalation and consumption of agricultural produce, but advice from MAFF suggests that this would increase the maximum dose by only about 10%. Doses to typical members of the fish eating public around Whitehaven are unlikely to exceed 10 μSv per year.

In addition to these recognised pathways, BNFL continues to monitor the edible seaweed *Porphyra Umbilicalis* although this is no longer harvested commercially from the Cumbrian coast for manufacture into laverbread. Nevertheless, even should the harvesting of Cumbrian seaweed and manufacture of laverbread resume, the dose arising would be expected to be less than 50 μSv per year.

The amounts of time spent by members of the public on intertidal areas of the coastline bordering the North East Irish Sea and more inland locations are also reviewed.[7,12] In West Cumbria, combining dose rates and occupancy times, MAFF identify a variety of groups (e.g. bait diggers, boat dwellers and fishermen) representative of those who receive the highest external exposures. Excluding the dose from other uptake pathways applicable to these people, the maximum exposure locally was 61 - 79 μSv in 1989.[12] An internal dose of about 100 - 120 μSv may be additive to the external component, arising mainly from shellfish consumption. External dose rates to houseboat dwellers on the Rivers Wyre and Ribble, in Lancashire, were estimated to have been up to 168 μSv in 1989[12] but additional exposure of these people due to consumption of fish and shellfish or handling of fishing gear is negligible.[7] BNFL measures dose rates over the mudbanks in Whitehaven harbour and elsewhere in order to assess a hypothetical critical group dose based on occupancy and seafood consumption rates applicable to the local fish-eating public. For this group the dose in 1990 would have been less than 150 μSv, exclusive of natural background.

Another hypothetical critical group is assumed to consist of people living close to the Sellafield site. The maximum dose incurred from the inhalation of particulate airborne material for this group is no more than

23 μSv and more typically would not exceed about 10 μSv. These estimates have declined somewhat over the past few years and assume continuous occupancy (i.e. 8 760 hours per year) at any location. This is unrealistic and any actual doses incurred will be lower than those estimated. In addition to activity on airborne particulates, the effects of activity present in gaseous or vapour form must be considered. These doses are estimated from discharge levels and, at the nearest habitation, would not exceed 75 μSv over the year due mainly to argon-41 from the Calder Hall reactors.

Based on advice from MAFF, the critical group for terrestial foodstuffs has always been assumed to be infants living at farms adjacent to the Sellafield site who obtain all their milk from those farms. The estimated dose to infants is less than 20 μSv and derives from caesium-137 and strontium-90; the concentration of most other radionuclides being consistently below the limits of detection. For adults the dose is no more than 10 μSv.

In addition to the sampling of milk, a limited programme of livestock and vegetable sampling is undertaken. In this case the highest dose estimate is for adults and would be between 10 and 45 μSv. In practice doses are lower, as the effects of buying food produced from a wide area are likely to reduce the actual dose received by up to a factor of ten.

If all the above terrestrial pathways (including inhalation and argon exposure) are considered to be additive, the maximum credible doses to adults is about 110 - 160 μSv. More typically the dose, even to critical group consumers, would be considerably lower.

Gamma dose rates at the Sellafield site perimeter are monitored routinely and average about 0.4 μSv per hour, although this does vary, particularly near the Calder Hall reactors. The increment above natural background level in the area is typically about 0.1 μSv per hour, which is due to direct radiation from the plant.

Members of the public spend very little time in the immediate vicinity of the Sellafield site perimeter so the measured radiation levels are thus of limited significance in terms of public radiation doses. The most exposed group of people identified is that occupying a nearby habitation where the annual dose, additional to natural background, is estimated to exceed 200 μSv. A specific analysis of pathways for this group suggests that an additional dose not exceeding 100 μSv may accrue from inhalation and food and water uptake. All this data on doses are summarised in Table 3.

Table 3 Estimated critical group doses to members of the public arising from Sellafield and Drigg environmental discharges in 1990.

Marine pathway doses	Terrestrial pathway doses
shellfish 110 - 130 μSv	meat 20 μSv
fish 15 - 30 μSv	milk 20 μSv
porphyra < 50 μSv	vegetable 20 μSv
occupancy* 150 μSv	water 10 μSv

Atmospheric pathway doses**
inhalation < 25 μSv
external 200 μSv

* - hypothetical pathway only.
** - considered to be an upper estimate. Studies are currently underway investigating this pathway.

5 COLLECTIVE DOSE ESTIMATES

Collective doses from discharges in the UK can be assessed against any defined population.[19,20,21,22] In practice, the dose is generally calculated for the UK or Europe. Collective dose commitments resulting from Sellafield discharges in 1989, summed over all time, were about 9 manSv for the UK population and 66 manSv for the European population including the UK.

An alternative presentation is used by MAFF, who evaluated the collective dose for the UK population in 1989, from all cumulative past discharges to sea from Sellafield, to be about 30 manSv.[7] This estimate has declined from 1976 when MAFF quoted a collective dose for the UK of 130 manSv, falling to 100 manSv in 1980 and 50 manSv in 1985.

Most of the collective dose commitment results from the discharge of carbon-14 because of its long radioactive half-life (5 760 years) and its incorporation into the global carbon cycle. The remainder of the collective dose commitment is attributable largely to iodine-129. However, individual doses from both these sources are generally small.

6 RISK ASSESSMENT

A linear dose-risk relationship is currently accepted, conservatively, by both the nuclear industry and its regulators, with no lower threshold for effect. Thus any radiation dose, including naturally occurring radiation, is presumed to give rise to some risk. Furthermore, the absolute risk per dose is assumed to be the same irrespective of the period over which it is delivered. This means that a single unit dose shared between a large population over 10 000 years carries an equal risk of a harmful effect arising as the same dose to a single person if delivered over one year. On occasions this can lead to the calculation of statistical risks for long-term collective doses which have little practical meaning. Nonetheless, the underlying principle of dose-risk linearity offers the best available model to ensure the safety of nuclear operations.

Recently the International Commission on Radiological Protection altered their estimate of risk factors,[23] which are now as follows:

fatal cancer	5% per Sv
non-fatal detriment (including hereditary effects)	5% per Sv

This gives a total risk of life impairment of 10% per Sv and includes the concept of "detriment" other than mortality. On this basis a dose of 500 μSv (the recommended annual limit for members of the public) presents a total health detriment risk of 1 in 20 000 or a fatal risk of 1 in 40 000. In practice the maximum sustainable exposure for any group is currently unlikely to exceed 200 μSv (Table 3), which presents a total health risk of 1 in 50 000 (or a fatal risk of 1 in 100 000) per year. Furthermore, long-term exposure at this level is unlikely for any individual. More typically, residents local to Sellafield but consuming a reasonable proportion of nationally grown produce (as available from most shops) are unlikely to receive more than 20 μSv per year additional to background. Further away, a typical person in the UK is unlikely to receive more than 2 μSv per year additional to background as a result of the entire UK nuclear power programme.[24] The risks incurred at these levels of exposure are proportionally lower.

A comparison of risks arising from a number of sources is presented in Table 4, from which it can be seen that, in relative terms, the impact of environmental discharges from nuclear operations is quite low. Furthermore, it must be recognised that, whilst some risk factors represent a short term or instantaneous effect and are therefore quite evident, the assumed risks of induced cancer from radioactivity generally appear only some 10 to 20 years after the causative exposure and are indistinguishable from those which may arise from background radiation or other causes. As a consequence the concept of loss of life expectancy has been developed.[26] Presenting risks in this way further emphasises the low impact of discharges from the nuclear industry (Table 5).

Finally, the theory of radiation hormesis, postulating that low-level radioactivity may have a positively beneficial effect on health and longevity, has gained some credance.[27] For the purposes of regulating the industry, this view has not been adopted and the well recorded lower mortality rates amongst UK nuclear industry workers[28,29,30] are generally ascribed to the "healthy worker effect".[31]

Table 4 Risk of an individual dying in any one year from various causes.

Source	Risk	Ref
smoking 10 cigarettes a day	1 in 200	25
all natural causes, age 40	1 in 850	25
any kind of violence or poisoning	1 in 3 300	25
influenza	1 in 5 000	25
accident on the road	1 in 8 000	25
radiation exposure at background levels (2 mSv)	1 in 10 000	23
playing soccer	1 in 25 000	25
accident at home	1 in 26 000	25
accident at work	1 in 43 500	25
homicide	1 in 100 000	25
radiation exposure at "critical group" level (0.2 mSv)	1 in 100 000	23
accident on railway	1 in 500 000	25
hit by lightning	1 in 10 000 000	25
typical radiation exposure from nuclear industry to UK population (0.002 mSv)	1 in 10 000 000	23

Table 5 Loss of life expectancy (LLE) from various causes[26] in days.

Cause	LLE (days)
being male rather than female	2 800
heart disease	2 100
being unmarried	2 000
cigarettes (1 packet per day)	1 600
cancer	980
stroke	520
15 lbs overweight	450
motor vehicle accidents	200
alcohol	130
small cars v. standard size	50
firearms	50
diet drinks (1 per day throughout life)	2
all electric power in US nuclear (calculated by the anti-nuclear Society of Concerned Scientists)	1.5
airline crashes	1
all electric power in US nuclear (calculated by Nuclear Regulatory Commission)	0.03

7 PUBLIC ACCEPTABILITY

The nuclear industry has the best safety record of any fuel industry in terms of health and longevity of its workforce. It is also among the least polluting of the major energy sources. Nonetheless, the industry remains controversial, fuelled partly by the continuing campaigns mounted by some elements of the media and by the high-profile nuclear accident at Chernobyl, USSR in 1986. Other accidents, including Windscale (1957) and Three Mile Island (1979) are also repeatedly referred to as "disasters" although no direct health effects within local populations have been identified. In addition, the many recent studies into possible links between leukaemia and low-level radiation[32] have contributed to the existence of a culturally induced "radiation phobia".[33]

The UK nuclear industry is currently subject to a Government review, with a policy statement for the future of nuclear power expected in 1994. It is thus vital to the industry that public support for its operations increases over the next few years. Clearly, appropriate presentations of the excellent technical and environmental performance of the nuclear industry must be cornerstones in gaining this support.

8 REFERENCES

1. "Radioactive Waste: The Government's Response to the Environment Committee's Report", CMND 9852, HMSO, London, 1986.
2. Radioactive Waste Management Advisory Committee, Fifth, Sixth & Seventh Annual Reports, HMSO, London, 1985, 1986, 1987.
3. National Radiological Protection Board, "Interim Guidance on the Implications of the Recent Revisions of Risk Estimates and the ICRP 1987 COMO Statement", NRPB-GS9, HMSO, London, 1987.
4. Her Majesty's Inspectorate of Pollution, "Monitoring of Radioactivity in the UK Environment", HMSO, London, 1987.
5. Department of the Environment Digest of Environmental Protection and Water Statistics, HMSO, London, 1987, no. 10.
6. Scottish Development Department, "Environmental Monitoring in Scotland: 1983 - 87", 1989, SDD 2(E).
7. Ministry of Agriculture, Fisheries and Food, "Radioactivity in surface and coastal waters of the British Isles, 1989", Aquatic Environmental Monitoring Report no.

23 Ministry of Agriculture, Fisheries and Food, Directorate of Fisheries Research, Lowestoft, in preparation.
8. Ministry of Agriculture, Fisheries and Food, Terrestrial Radioactivity Monitoring Programme (TRAMP), "Radioactivity in food and agricultural products in England and Wales", MAFF publications, Alnwick, Report for 1989, no. 4.
9. W.A. McKay and J. Harold, "A study of silt and shellfish radioactivity levels in Cumbrian near-shore waters, 1987", AERE-R 13192, Harwell, 1988.
10. Isle of Man Government Laboratory, "Environmental radioactivity monitoring on the Isle of Man, 1989", 1990.
11. G.J. Hunt and J.G. Shephard, "The identification of critical groups", Fifth International Congress of IRPS, Jerusalem, 1980, **III**, 149 - 152.
12. T.C. Doddington, W.C. Camplin and P. Caldwell, "Investigation of external radiation exposure pathways in the Eastern Irish Sea, 1989", Ministry of Agriculture, Fisheries and Food, Lowestoft, Fisheries Research Data no. 22, 1990.
13. T.H. Stewart, M.J. Fulker and S.R. Jones, *J. Soc. Radiol. Prot.*, 1990, **10**, 115 - 122.
14. National Radiological Protection Board, "Dosimetric Quantities and Basic Data for the Evaluation of Generalised Derived Limits", NRPB-DL3, HMSO, London, 1980.
15. National Radiological Protection Board, "Committed Doses to Selected Organs and Committed Effective Doses from Intakes of Radionuclides", NRPB-GS7, HMSO, London, 1987.
16. National Radiological Protection Board, "Revised Generalised Derived Limits for radioisotopes of strontium, iodine, caesium, plutonium, americium and curium", NRPB-GS8, HMSO, London, 1987.
17. International Commission on Radiological Protection, ICRP Publication no. 26, Pergamon Press, 1977.
18. D. Jackson, S. Bradley, M.R. Hadwin and P. Dockrell, "The effect of soaking on radionuclide concentrations in winkles from the Cumbrian coast", in Report of the 9th annual meeting of NERC/COGER, Egham, 1990.
19. National Radiological Protection Board/Commissariat à L'Energic Atomique, "Methodology for Evaluating the Radiological Consequences of Radioactive Effluents Released in Normal Operations", CEC V/3865/79, Luxembourg, 1979.
20. W.C. Camplin and M. Broomfield, "Collective Dose to the European Community from Nuclear Industry Effluents Discharged in 1978", NRPB-R141, HMSO, London, 1983.
21. International Atomic Energy Agency, "A practical methodology for the assessment of collective radiation doses from radionuclides in the environment",Draft working document, 1987.
22. G. Lawson, J.R. Cooper and N.P. McColl,"The Radiological Impact of Routine Discharges from the UK Civil Nuclear Sites", NRPB-R231, HMSO, London, 1988.

23. International Commission on Radiological Protection, ICRP Publication no. 60, Pergamon Press, 1990.
24. National Radiological Protection Board, "Living with Radiation", 4th edition, HMSO, London, 1989.
25. British Medical Association, "Living with Risk", Wiley, 1987.
26. B.L. Cohen, "Before it's too late: a scientist's case for nuclear energy", Plenum Press, New York, 1983.
27. J.H. Fremlin, *Atom.*, 1989,**390**, 4 - 7.
28. E.A. Clough, *J. Soc. Rad. Prot.*, 1983, **3**(i), 24 - 27.
29. P.G. Smith and A.J. Douglas, "Cancer mortality among workers at the Sellafield plant of British Nuclear Fuels", in "Health Effects of Low Dose Ionising Radiation", BNES, 1988, 71 - 76.
30. E.A. Clough, *J. Soc. Rad. Prot.*, 1983, **3**(3), 18 - 20.
31. V. Beral, L. Carpenter, M. Booth, H. Inskip and A. Brown, "The 'healthy worker effect' and other determinants of mortality in workers in the nuclear industry", in "Health Effects of Low Dose Ionising Radiation", BNES, 1988, 95 - 100.
32. M.J. Gardner, M.P. Snee, A.J. Hall, C.A. Powell, S. Downes and J.D. Terrell, *Br. Med. J.*, 1990, **300**, 423 - 429.
33. W.R. Hendee, *Health Phys.*, 1990, **59**, 763 - 764.

THE ROLE AND EXPERIENCES OF AN ENVIRONMENTAL CONSULTANT

P. J. Young

Aspinwall & Company Ltd
Walford Manor
Baschurch
Shrewsbury
Shropshire, SY4 2HH
United Kingdom

1 INTRODUCTION

Recently there has been a rapid growth in public awareness and concern over environmental issues, ranging from global warming to waste recycling. This has led to the advent of green consumerism and an increase in the activity of environmental pressure groups. Media attention has also focused on environmental problems and many companies have consequently become very sensitive to their public image. The chemical industry has received considerable attention as a major polluter and therefore finds itself at the forefront of the call for improved environmental standards.

This enhanced environmental awareness is impacting on board room policies and on politics, such that there is considerable pressure on industry to adopt a more responsible environmental policy and on regulators to ensure that it does so. The Water Act 1989 and the Environmental Protection Act 1990 (EPA) have been introduced into this climate as the mainstay of government legislation to demonstrate that environmental issues are high on the agenda. An ever broadening range of environmental regulation is also being introduced within the European Community (EC) to provide much of the momentum to individual national legislative programmes.

Consultants can have much to offer to help to achieve good environmental management in industry. Two criteria are paramount in qualifying the environmental consultant to make a valuable contribution:

a) Expertise - an ability to apply specialist knowledge and experience to augment the capabilities of the client, whether regulator or industry.

b) Independence - an objective approach which will stand up to vigorous technical, legal and moral scrutiny, to help build consistency and confidence in environmental decision making.

From these key points flow all the economic and practical benefits of utilising a consultant's know-how where employed resources are overstretched.

This paper reviews the role of a consultant in environmental affairs, including how this role is seen by both the client and the consultant. The capabilities which may be called upon are outlined, with particular reference to the field of air pollution. Five case studies involving the chemical industry have been chosen as a way of demonstrating practical examples of the contribution that the environmental consultant can make.

2 THE ROLE OF THE CONSULTANT

From the consultant's viewpoint an increasingly important feature of good environmental advice is the ability to think across many disparate disciplines. This capability is becoming more widely recognised following the introduction of the concept of integrated pollution control (IPC) in Part 1 of the EPA. The main objectives of IPC are stated [1] to be:

a) To prevent or minimise the release of prescribed substances and to render harmless any such substances which are released.

b) To develop an approach to pollution that considers discharges from industrial processes to all media in the context of the effect on the environment as a whole.

There has been much debate as to the ability of Her Majesty's Inspectorate of Pollution (HMIP) and the National Rivers Authority (NRA) in England and Wales, who are responsible for administrating IPC, to be able to take a sufficiently balanced view of the impacts of prescibed substances released to different media. The application of IPC will therefore provide a pertinent opportunity for consultants to assist in applying the principles of "best available techniques not entailing excessive cost" (BATNEEC), to demonstrate that the impact of industry on the environment can be genuinely minimised by proper assessment of air, land and water quality. Indeed the whole ethos behind the EPA significantly shifts the protection of the environment from a piecemeal exercise requiring consultants with specific expertise in narrow specialised areas, to a situtation where a holistic approach is essential.

There will always be examples of individual technical problems where consultancy assistance has a role; this may be met by the services of appropriately qualified individuals or specialised organisations. However, increasingly the context of individual problem solving will become important, with the consequences of actions in such as process design or abatement technology needing understanding within a wider context. From a consultant's viewpoint it is therefore important to appreciate the full range of activities at a chemical plant before informed advice on BATNEEC or general good practice can be offered.

Consultants will only benefit the interests of their client (whether industry or a statutory authority), the environment and the public, if they provide independent and objective advice and make "care of the environment" one of their priorities. Giving clients the answer they would rather hear does not act in anybody's interest and indeed could put the client at a disadvantage in the long-term. Thus consultants have a very important role to play in educating all parties to adopt consistent priorities and standards, and in prompting the sharing of relevant experience.

In industry, or indeed for any client, probably the most important criteria for the successful appointment of an environmental consultant is confidence that successful integration of the consultant can take place with the in-house technical and management teams. This problem can be compounded where the environmental know-how available within the client organisation is limited, since it becomes harder to judge and prove that the consultant is genuinely capable of delivering the expertise and advice sought. It is believed that the proliferation of environmental consultancies during the latter part of the 1980's benefited as much from the naivety of clients seeking environmental advice as it did from the genuine increase in demand for environmental work.

There are many signs that the chemical industry now gives the environment a high profile within the board room and management structure. Such initiatives as the draft "British Standard" on environmental management systems, recently published for consultation by the British Standards Institution,[2] demonstrate the prudence of giving environmental issues close attention in the industry. However, it is clear that British industry still has reacted slowly to environmental pressures. For example, a recent survey by the Institute of Directors[3] has shown that 77% of company directors claim they have no formal corporate environmental policy and only 4% of directors claim to have good knowledge of existing environmental legislation.

A second crucial element uppermost in the mind of someone appointing a consultant is the ability to ensure cost effectiveness in the work to be carried out. Again this requires good understanding of the services being offered if a judgement is to be made on a basis of value for money and not, as so often happens, by choosing the cheapest tender. Indeed where a detailed specification or project brief has not been prepared, it is likely that wildly different packages will be offered to the bewildered client, reflecting the wide ranging interpretations of the client's needs by different consultancies. Where resources permit, and many chemical companies have taken this route, it is clear that a properly trained environmental manager will greatly aid the selection of, and liaison with, a good consultant. In the absence of such resources it is strongly advised that the experiences of colleagues from other similar organisations be sought to identify a consultant who can be appointed purely to scope and define the brief where significant investment is being considered.

The attitude of regulatory authorities to consultants is also of great importance. Increasingly organisations such as the NRA and HMIP are themselves augmenting in-house expertise by employing consultants. They are therefore familiar with the advantages and pitfalls of using consultancy advice. Private sector clients often look to consultants to provide an element of independence and enhanced credibility with regulatory authorities to secure the smooth progress of environmental or development projects. The standing which a consultant enjoys with respect to the regulator is therefore very important whether the consultancy is working in the public or private sectors.

A further sector of relevance is wider public opinion, including local and even national politicians. Any project is likely to receive environmental scrutiny in a public forum since environmental issues are high on the political agenda, the local government decision making process is based on committees of elected members and environmental pressure groups have become much more active. The role of the environment consultant must therefore address not just the technical and professional advice, but also provide effective communication and presentation of issues in a non-technical way. This represents a particular challenge given that there is a broad consensus that the scientists and engineers, who are so often at the centre of environmental work, have been poor communicators in the past.

3 THE TECHNICAL CAPABILITIES NEEDED

It follows from the increasingly broad base of environmental work that individuals from many different professional backgrounds have been drawn into the environmental field. It is not intended to list the skills of relevance in this paper, but as an example to give a brief description of the key components of expertise in one field, that of air pollution. Before considering this example it is worth highlighting the benefits of using available reference guides[4] which are capable of listing the disciplines and subject areas of individual consultancies. This information is valuable in reflecting where the strengths and weaknesses of a consultancy lie and in ensuring that there are the back-up resources to provide well balanced advice. Where environmental consultancy is a subsiduary activity, care is needed to ensure that the advice is, and will be perceived as, unbiased by the main interests of the organisation. For example, solutions are often advocated which are solely on the products which are available within the parent organisation to which the consultancy belongs. The formation of such groups as the Association of Environmental Consultants will also help in the selection of appropriate expertise.

In areas such as environmental auditing, environmental assessment and IPC, a wide range of disciplines will be called upon even within relatively small projects. However, where advice is sought within a more specific area such as contaminated land,groundwater protection or air pollution a specific professional background will dominate. Even so a range of techniques will need to be applied in the office and on-site to solve the problem.

Taking the example of the provision of technical assistance in air pollution, two scenarios are likely to represent the project. Firstly, the object may be to protect air quality by identifying in advance the likely impacts of a newly proposed development or activity. Secondly, an improvement in existing air quality may be sought through the assessment and mitigation of existing emissions. In both cases a rigorous approach by the consultant is necessary to give technically and legally defensible information of sufficient quality to satisfy the needs of the client and/or the regulating authority. Four key technical areas have been identified which make up a full air quality impact study for a given process; these demonstrate the breadth of expertise needed to tackle air pollution issues.

4 AIR POLLUTION CONTROL - FOUR KEY STEPS

1. Baseline Surveys (new developments) or Audits (existing processes)
2. Monitoring Programme
3. Computer Modelling
4. Pollution Abatement Technology

Baseline Survey

In a baseline survey an assessment is carried out on a proposed development:

a) Existing and potential air quality criteria are defined using a full understanding of the materials and processes involved in the proposed development which may lead to atmospheric emissions.

b) A review of relevant legislation is carried out and clarification from regulatory authorities sought.

c) A priority list is drawn up for substances of concern which forms the basis of the baseline survey and any predictive modelling.

d) As part of the impact study the prevailing background pollution levels need to be established, against which future contributions attributable to a development may be measured. The existing levels of the selected gaseous and particulate sudstances are thus determined in order to assess ambient air quality.

Environmental Auditing

Environmental auditing is a management tool for assessing and reporting on standards of environmental performance of an existing facility. It provides a rigorous, systematic and effective means of assessment and therefore is becoming an important part of environmental quality control and management for many companies. In particular, an audit can provide information to managers about how well an industry's operations and activities comply with company targets, standards and codes of practice, as well as the relevant legislation. Auditing can greatly benefit from a consultant's independence and external experience to put the current performance of a plant in the context of its competitors and industry standards. The key information from the audit should identify:

a) Baseline information about local environmental conditions and activities.

b) Strengths, weaknesses, opportunities and threats in relation to the quality of the local environment and activities/performance of the organisation being audited.

c) Recommendations for environmental policies, management strategies and action plans.

d) With respect to air quality and air emissions, relate these to the wider context of all environmental media (an essential first step for the application of IPC).

The baseline survey or audit will make use of data already available. Where deficiencies are identified, a *monitoring programme* must be defined. An air monitoring programme will be centred around target receptor locations, chosen because of their sensitivity to particular substances and taking into account likely pollutant pathways. Other potential sources of pollution also need to be considered if situated in the vicinity.

Monitoring is often under-estimated and as a result the data collected may not fulfil even the simplest needs. Appropriate consulting support can ensure that monitoring is targeted and implemented successfully and cost effectively. An indication of the importance of monitoring is the draft guidance recently put out for consultation by HMIP for inspectors defining monitoring for processes subject to IPC. The key elements of good monitoring are expected to be identified in the guidance when it is published and are summarised in the box.

With many processes authorised under Part 1 of the EPA being subject to specific and onerous monitoring requirements, the prudence of putting resources into a correct approach to monitoring is clear. For all but the most simple IPC applications, monitoring requirements will be specified in a supporting document separate from the authorisation. Both regulator and industry will benefit if the expertise is available for the operator to draft this document on the basis of experience gained at the time of the application. Increasing opportunities to use on-line monitoring instruments linked to computerised data storage systems will undoubtedly be exploited by regulators to provide the best quality data.

From experience good consultants can play a very helpful role in achieving monitoring objectives and can provide the short term resources to implement once-off or irregular monitoring needs, as well as implementing and training staff in routine procedures. Of particular help is often the provision of customised software to store, manipulate and prompt action from the data obtained.

Measurement of emissions at source does not reflect the fate of pollutants once they leave the site, so when the information on emissions has been obtained, **computer modelling** may be performed to indicate the destination of the pollutants within the local environment. However, sampling of the atmosphere is preferable to obtain a more realistic impression of what is actually happening, in which case a baseline survey will be needed to put into context any contribution from the site to neighbourhood pollution levels

Air Monitoring	
Emission Points	The exact location of monitoring must be specified, both for point sources such as stacks and for fugitive emissions.
Monitoring Methods	The selection of the correct determinands for monitoring must first be made and agreed. The suitability of continuous on-line measurements or equipment for extracting spot samples must be assessed. The standards or methodologies for sampling and analysis require agreement and with many emissions not falling within existing British Standards, reviews of international standards or even drafting of new methodologies may be needed. Instrumentation availability and calibration require definition.
Monitoring Frequency	Where continuous on-line monitoring is not practical, the frequency of intermittent sampling and analysis will require justification.
Laboratory Analysis	Where laboratory analysis is required, satisfactory quality control and assurance procedures must be demonstrated and an appropriate level of inter-laboratory comparison or duplicate testing may be necessary.
Record Keeping	The method, frequency and storage of monitoring data, including any requirements for submission to the regulatory authority, will require prior agreement.

Following characterisation of the pollution sources, a number of computer dispersion models may be used to predict exposure concentrations at the sensitive receptor locations and to highlight whether additional risk zones exist by production of pollution contours over the study area. Models typically used are point sources for chimney stack emissions or sources for fugitive emissions.

The significance of the predicted air pollutant concentrations is evaluated by direct comparison with existing or proposed legislative standards for ambient air. In their absence, comparison can be made with occupational exposure standards (appropriately weighted to extend their use to the general public) or to standards being imposed abroad. The standard to be adopted is often the subject of involved negotiation between the regulator and the industry if the one adopted may have a significant influence on whether or not costly abatement equipment is needed.

The unacceptability of the predicted impact will be a major factor in deciding whether **pollution abatement technology** is to be considered as part of the development strategy. Experienced consultants can advise on the suitability of various abatement technologies, the anticipated effectiveness of which can be predicted by dispersion modelling.

Predictive modelling can also be used to determine acceptable criteria for "dilute and disperse", should this be the chosen option for mitigation, for example by defining chimney height and efflux velocity.

The issue of abatement technology in the context of the EPA immediately raises the concept of BATNEEC. It is important to recognise the unusual legal position which industry finds BATNEEC imposing. In the United Kingdom the law normally aims to prosecute to beyond reasonable doubt; the accused is presumed innocent until shown guilty. For prescribed processes the onus of proof lies with the defence to demonstrate BATNEEC; the presumption of the regulator is that pollution is unacceptable until convinced that BATNEEC has been applied. Consultants have a role for industry in helping them fulfil this unusually demanding position, particularly when BATNEEC guidance notes are not available in a final form or where local circumstances dictate application of more than the minimum standards.

A second key issue is the integrated approach required to demonstrate BATNEEC being applied; both in terms of having multi-disciplinary expertise to argue effectively about the comparative merits of discharges to air, land and water, and in terms of relating the environmental effects of the emission to the cost and technology of the abatement method. Industry usually has a clear view of its product, plant operation and waste disposal costs, but a poor understanding of how to evaluate environmental impact or the costs of disposal of new waste streams generated by process changes or new abatement techniques. A very large scale example of this has been the massive economic adjustments in the costs of flue gas desulphurisation of major power stations as the real value of the by-product gypsum waste became apparent.

The application of BATNEEC therefore raises unusual difficulties, particularly in the absence of case law and precedents to guide interpretation. Useful consulting experience in evaluating environmental benefits, accurately predicting pollution control costs and providing genuinely balanced views on cross media options for IPC, is at a premium. However, the vital words are "useful experience" as many consultants claim this, such expertise appears much more widespread than is the case, and the validity of such claims must be tested before appointment to ensure sound advice is on offer.

5 CASE STUDIES

A selection of five case studies have been chosen based on commissions carried out recently by Aspinwall & Company and which relate to the environmental impact of the chemical and related manufacturing industries. These case studies broadly represent advice in the areas of air emissions, environmental auditing, environmental assessment, contaminated land and environmental policy advice. These five subject areas represent perhaps the commonest issues where the chemical industry has taken consultancy advice.

Case Study 1 - Air Pollution

A company intended to construct a new manufacturing plant involving processes liable to give rise to air and dust emissions. Three distinct tasks were identified in order to provide the necessary advice and information in support of the development application. These were a baseline study, air pollution modelling and a dust impact study.

The location of the site overlooking a steep sided valley with several residential communities represented a relatively complex local environment in which to predict air emission impacts. A three week monitoring programme was carried out to establish existing levels of sulphur dioxide, nitrogen dioxide, fluoride and chloride to provide baseline information. These parameters were the air pollutants identified in the current draft BATNEEC notes for the industry. Good correlation was obtained with existing background data for the area and confirmed the long term monitoring record which indicated relatively good air quality in the vicinity of the proposed site.

The monitoring data was then used to carry out short and long term computer modelling for all four parameters to predict the possible impact of air emissions from the main stack on four sensitive residential locations. The short term modelling exercise used worst case meteorological conditions. These had been identified by a combination of site specific data from a portable meteorological station erected on-site and the local Meteorological Office weather station 24 km away. The background concentrations were added to the ground level concentrations predicted by the point source dispersion model to obtain maximum concentrations in two ways.

For sulphur dioxide the highest twenty four hour concentration measured over a recent three year period was added to the maximum modelled ground level concentration to indicate the worst daily case. This value was compared with the statutory yearly EC limit value of 0.35 mg m^{-3}.[5] From this it was deduced that there was no problem, since the predicted concentration lay below the yearly mean statutory level and the predicted yearly mean will clearly be much lower than this worst case. A comparison was also made with the stricter, shorter term, non-statutory twenty four hour EC guide value. For nitrogen dioxide, fluoride and chloride the maximum predicted concentrations were added to the maximum from the three week baseline study. Nitrogen dioxide concentrations were compared with the yearly EC limit and guide values[6] and the short term World Health Organisation twenty four hour guideline concentration.[7] For chloride and fluoride air quality criteria were derived by comparison with the occupational exposure standards published by the Health & Safety Executive[8] divided by a safety factor of 80. This approach provided a rigorous confirmation that in terms of human health the stack emissions were unlikely to have any significant impact on air quality in the valley.

Work was also carried out which demonstrated from a similar plant that those particulates released from the process were dominated by smoke particles which could be adequately predicted by dispersion modelling. Subsequent modelling demonstrated that particulate emissions would not have a significant impact on the surrounding area. An assessment of odour releases was also made using the same modelling approach and no odour problem could be predicted.

Monitoring of ambient dust levels at the sensitive receptors as well as the control location was performed on four occasions over a two month period and all measured concentrations were found to be below the derived criterion for air quality for total inhalable dust. Existing activities on-site were reviewed such that current best practices could be shown to be employed to mitigate dust nuisance and an assessment was made to show that the new plant would not give rise to any noticeable deterioration of the present circumstances. Finally, a limited assessment was made of fluoride impacts on surrounding soil and vegetation by sampling and analysing their fluoride content to provide background data. Future fluoride concentrations can be compared with these background data to demonstrate that the prediction of no impact is justified.

Case Study 2 - Waste Management Audit

A major integrated chemical manufacturing site was selected by the client for a waste management audit. The objective was to provide an information database on waste arisings and disposal activities and to assess this in the context of existing legislation, the provisions of the Duty of Care as defined in the EPA and best economic practice. Waste management procedures, waste handling and waste disposal facilities were all audited during an intensive two month period.

A standardised set of questionnaireswas used to gather compatible data from all the process plants and management functions with waste arisings and waste management responsibilities. A customised personal computer database of all waste movements was compiled and interrogated to provide the first detailed and comprehensive source of information on waste movements for the client. The effectiveness of the audit was greatly helped by the resources committed to it by the client. The audits were effectively carried out by a team consisting of the client's own personnel working alongside the auditors from Aspinwall & Company.

A total of 223 waste streams were identified with 18% of the controlled wastes being removed for recovery or recycling. Special wastes represented 32% of the total, a high proportion which is typical for the chemical industry and one which is likely to rise rather than fall as total waste arisings are minimised by more efficient process design and control. Sixteen licenced storage and disposal facilities were audited.

The audit revealed a wide range of deficiencies in the implementation of waste management responsibilities. Indeed this experience is not unusual and considerable effort will be needed by industry if the Duty of Care requirements are to be met, given the low priority waste management has received in the past. This client was typical, in Aspinwall & Company's experience, in that disposals likely to contravene site licences were identified, no policy for waste disposal existed at the time of the audit and internal management guides and policy were often flouted. Recommendations for improvements at all levels of the management structure were made under five key headings: Waste Information Issues; Waste Management Policy Issues; Waste Management Procedures; Appraisal of External Contractors; and Waste Documentation.

One feature common to all waste management audits of this type carried out by Aspinwall & Company is the low priority given to the control of skips on site and to the storage and disposal of drummed waste. These two issues seem to be ubiquitous in the industry, and, whilst many good examples can be quoted, it is suggested that few sites can clearly demonstrate that every skip consigned for disposal has wholly known contents and that in no quiet corner lurk a few drums representing a current environmental liability and a future waste disposal problem. Finally, the scope for reducing waste disposal costs by better handling, storage and control of wastes almost always outweighs the additional costs of providing more environmentally sound disposal of those waste streams where current practice is found wanting.

Case Study 3 - Environmental Assessment

For this example an occasion where the role of Aspinwall & Company was provision of advice to the regulator has been chosen. Outline and subsequently detailed planning applications were received for the proposed construction of an organic chemical synthesis plant on an undeveloped, although contaminated, site covering 16 hectares. The client, as planning authority, requested advice to assess the applications from a technical and planning viewpoint.

The project was characterised by a full and open dialogue with the applicant which greatly assisted in overcoming concerns before and after the environmental statements in support of the planning applications were submitted. A full understanding of the proposed development was used to provide a non-technical overview for the client and to establish control requirements of the regulators. On behalf of the client, liaison with both statutory and non-statutory consultees was completed on an accelerated timescale and agreement reached on key issues with HMIP, NRA, the local water plc, the Health & Safety Executive and the local authority environmental health department.

A report detailing the nature of the chemical processes likely to be carried out was produced which considered the impact of any emissions to air and water, together with the control of potential accidental discharges and spillages on site. The applicant was encouraged to develop the proposals with respect to the investigation and remedial designs for existing on-site contamination, the provision of clear emergency plans including evacuation procedures and the scope of environmental monitoring. On reaching the point where planning permission could be recommended, advice was provided on conditions to be included in the planning permission and on heads of terms for legal agreement between the applicant and the planning authority.

Whether acting on behalf of applicant or planning authority, the need to provide a high quality environmental assessment is apparent if the best conditions for approving development are to be created. This requires the input of specialists to each relevant environmental impact identified by an initial scoping exercise. In addition the management of such projects is preferably held by an experienced environmental planner as it is the purposes of planning legislation which must be fulfilled. Experience in dealing with complex applications such as a new chemical plant is particularly apparent and beneficial when negotiating conditions for planning permission and IPC and in the drafting of legal agreements which are almost inevitable in developments of this type.

Case Study 4 - Contaminated Land Assessment/Remediation

This area of work, encompassing surface and groundwater contamination surveys, represents the most frequent environmental consultancy brief received by Aspinwall & Company from chemical industry clients in recent years. The example chosen is made up from components of two briefs for separate sites to indicate how the whole process from initial investigation through to remediation has been tackled.

The first site involved a chemical works with adjacent contaminated land formerly used for disposal of process wastes covering over ten hectares. Some manufacturing activity continues, but much of the site has been razed to the ground which is covered in old concrete pads, foundations and pipework reflecting over 100 years of manufacturing activity. As a first step a detailed site history had been compiled by the client and this was corroborated and enhanced by researching records from local history sources, anecdotal information from employees, aerial photographs and early legal and consent documents. Using this information, together with published geological records, a preliminary contaminated land survey involving hydrogeological boreholes, trial pits and soil vapour probing was designed. The analytical programme was also outlined, based on the anticipated contaminants in the site and the sensitivity of the location.

Site work was carried out according to strict health and safety procedures and with minimum disruption to manufacturing activity. Several buried structures were encountered but these did not significantly alter the scope of the investigation. Detailed supervision and logging of all excavation and borehole drilling enabled laboratory analytical work to be targetted effectively using cascade analytical suites, whereby high levels of qualitative measurements trigger increasingly detailed quanitative analysis. For example, soil vapour surveys may indicate where sampling for organic contamination is needed; high total solvent extracts then show which samples need specific measurements for polyaromatic hydrocarbons, or where chlorinated compounds are identified, for polychlorinated biphenyls. Following completion of the hydrogeological boreholes, which confirmed contamination of a relatively shallow and uniform aquifer, a period of twelve months monitoring to detailed seasonal groundwater flow and quality ensued. Specific intensive investigations of small sensitive areas were also carried out in the light of the initial results.

Interpretation of the soil and water analyses requires prudent use of published standards; those offered by the Interdepartmental Committee on the Redevelopment of Contaminated Land[9] and the Building Research Establishment[10] in the United Kingdom and the Dutch "ABC" standards[11] are the most useful for soils. However, standards used in different countries do vary and in many urban areas contaminants such as lead and arsenic are endemic. Groundwater and drinking water legislation, most notably that of the European Community, will assist in assessing groundwater contamination, although local background water quality will be important. Once a three dimensional picture of contamination is available and the mobility of contaminants is deduced from their solubility, infiltration rates and the hydrogeology, remedial measures can be assessed and costed.

To complete this case study a similar site to that described above has reached the stage following a review of the options for remedial work. The client is now being advised on the implication of the contamination for redevelopment and reclamation of the site, and engineering specifications have been drawn up for remedial works. Reclamation will be achieved through selection of a redevelopment partner who will assist in bringing optimum commercial value to the land, whilst ensuring that contamination problems are solved and do not represent a liability for the future. In this case much of the contamination is inorganic and relatively immobile so that localised treatment with encapsulation will provide the best solution. Indeed it is worth highlighting that although many new treatment technologies including vitrification, chemical fixation, biotechnology, air stripping and groundwater treatment have been evaluated for clients, very few indeed have yet implemented solutions other than removal for licenced disposal or encapsulation/capping. Nor have the terms of the government grant aid to date greatly encouraged a more permanent, but more expensive, clean-up technology to be chosen.

Case Study 5 - Environmental Policy

Consultancy advice on environmental policy embraces the full breath of industry's relationship with the environment and the ultimate users of its products. At one extreme Aspinwall & Company has advised the Department of Trade & Industry on the implications of existing and likely future United Kingdom environmental legislation on the chemical industry, in comparison with other industrial sectors, but the example chosen is a more specific brief relating to a new coating technology. The client required strategic advice on the environmental suitability of a new coating technology with a specified market application. A full health, safety and risk assessment was neither required nor appropriate as the product was still under development. However, when authorising often expensive research and development programmes it is important to identify at the earliest stage possible the environmental strengths and limitations of a new product or process. In extreme cases, for example replacements for chlorofluorocarbons, the environmental performance will be paramount.

This study therefore involved a detailed review of worldwide legislation enacted or predicted, which would impinge on the use and application of the new product. The most important hazards arising from the chemicals used in the process were identified together with sensitive applications requiring the most stringent standards. The review was carried out in two parts, the first dealing with occupational health impacts and the second with environmental impacts.

The occupational health study concluded that appropriate control measures could be applied to meet international standards without any serious economic disadvantage to the new coating technology. The most sensitive components were identified so that future development work would be cognisant of the likely impacts of reducing or increasing their contribution to the product formula.

The environmental impacts showed that the initial washings following the application of the coating were likely to be subject to stringent control in most countries. Recommendations were therefore made to restrict this activity to locations where appropriate control and treatment of discharges of the washings can be demonstrated. It was considered that the washings would lend themselves to the development of recycling methods to minimise the ultimate discharge and hence reduce the cost of providing efficient collection and treatment. Given that good housekeeping will avoid all potential environmental problems and that treatment and control technologies already exist for the contaminants of concern, successful development of the product is unlikely to be limited by poor environmental constraints.

6 REFERENCES

1. Department of the Environment/Welsh Office, "Integrated Pollution Control: A Practical Guide", HMSO, 1990.
2. British Standards Institute, "Draft British Standard on Environmental Management Systems EPC/50", June 1991.
3. Institute of Directors, "Taylor Nelson Survey", August 1991.
4. Environmental Data Services Ltd, "Directory of Environmental Consultants", second edition, 1990.
5. European Council Directive 80/779/EEC, "Air Quality Limit Values and Guide Values for Sulphur Dioxide and Suspended Particulates", 15 July 1980.
6. European Council Directive 85/203/EEC, "Air Quality Standards for Nitrogen Dioxide", 27 March 1985.
7. World Health Organisation, "Air Quality Guidelines for Europe", European Series No. 23, 1987.
8. Health & Safety Executive Guidance Note EH40/91, "Occupational Exposure Limits", HMSO, 1991.
9. Department of the Environment, "Guidance on the Assessment and Redevelopment of Contaminated Land", ICRCL 59/83, second edition, July 1987.
10. Building Research Establishment, "Concrete in Sulphate Bearing Soils and Groundwater", BRE Digest 250, 1986.
11. National Institute of Public Health and Environmental Protection, The Netherlands Report No. 738708002, "European Experience in Hydrocarbon Contaminated Groundwater and Soil Remediation", January 1989.

RESPONSIBLE CARE OF THE ENVIRONMENT:
A VIEW FROM INDUSTRY

R J Hulse

BASF plc
Seal Sands
Middlesbrough
Cleveland, TS2 1TX
United Kingdom

1 INTRODUCTION

To illustrate a view from industry a description of the experiences of BASF's approach to the Chemical Industry Association's initiative called "Responsible Care" is outlined in this paper. Also covered is the background to the Chemical Industries involvement with "Responsible Care", the response of BASF to the challenges of the programme and the management systems which were then adopted. Another initiative of the Chemical Industries Association (CIA) called "Speak out and listen" is also described.

In conclusion an illustration of Responsible Care in Action is given by describing briefly a major incident involving iso tank containers washed overboard from a North Sea Ferry and the subsequent effects when two of these tankers came ashore on a North Norfolk beach.

2 BACKGROUND

The Responsible Care programme originated in Canada in the mid 1980's. Its idea of opening up the black box of the chemical industry's sites to the community surrounding those locations, spread across the border into the USA a year or two later.

By 1989 it had reached the United Kingdom (UK) via the Chemical Industry Association, who realised that something was required to reverse the disastrous trend of rapidly falling public opinion of the industry for which a Mori survey showed that only one industry was lower in the polls, namely the Nuclear industry, largely in the wake of the Chernobyl disaster.

A task force was established under the Chairmanship of Dr Jim Whiston of ICI and from the outset it was agreed that the Resonsible Care Programme must amount to more than just a public relations exercise - it must demonstrate a new era of openness about the Chemical Industry and importantly, that it must show a commitment to a genuine improvement in performance.

3 BASF AND RESPONSIBLE CARE

In the mid 1980's BASF, the largest chemical company in the world, began a process of acquisition in the UK and changed its profile in the UK from predominantly an importer and distributor of German products, to one of a major manufacturer, as diverse as a major petrochemicals plant on Tees-side, to a number of small coatings and inks plants. The geographical spread was away from a concentration in the North West to plants, warehouses and offices from Scotland to Sussex.

A survey of these sites in 1989 established that BASF had a wide spread of standards in its operations ranging from near outstanding to potentially major problems in older locations. After an intensive study it was decided to concentrate on manufacturing and warehousing operations in the UK in order to establish uniform standards in Environment, Health and

Safety at all locations, including administrative offices and warehouses.

4 ESTABLISHING AN ENVIRONMENT HEALTH AND SAFETY POLICY

The board of BASF plc embarked, in 1989, on a new initiative and produced a Policy statement (Appendix 1). A new Health and Safety Policy was established alongside a new Environment Policy to show the importance to the Group of the emerging public awareness of the importance of the Environment to the continuing success of all new businesses in the UK in the 1990's.

To unite the many diverse operations a common policy was directed and to demonstrate the commitment from the top of the Company, the Managing Director and the Site Director from each location were signatories on each policy statement at all UK sites. This document also showed that the Group was to appoint a new senior executive to be responsible for the implementation of some uniform standards in Environment, Health and Safety and also to state clearly that each location head was in turn responsible for the preparation of detailed procedures at their site.

This was regarded as a fundamental requirement of the implementation of our policy and was a step change at some of the older established locations. Here they had grown up expecting the old safety departments to write their procedures and generally to be responsible for all Health and Safety issues. The newest acquisition on Teesside had a long and successful track record in Environment, Health and Safety inherited from its former American owners. The board quickly realised that the high standards of this location were likely to be the best available source in the UK on which to build uniform standards for the rest of the UK operations.

5 TRANSFORMING POLICY INTO PRACTICE

The process of transforming *policy* into *procedures* was started by the formation of a new UK Group of existing professional managers from the established Safety Managers and the new Occupational Hygiene and Environmental Specialists who became the working group charged with defining a set of Guidelines for the UK group (Appendix 2).

These draft guidelines were restricted to simple one or two page documents covering the most important aspects of Safety, Health and Environmental issues. However, to ensure acceptance by the line managers, a new Group Environment, Health and Safety Steering Committee was formed to approve these guidelines and later in 1990 to form Audit teams to assess each location in relation to these new standards.

6 THE ENVIRONMENT, HEALTH AND SAFETY AUDIT

Starting in October 1990 and continuing at the rate of about two locations per month, all the major locations in the UK were *audited* by March 1991. It was important in this process that mixed teams of senior managers from each location, supported by Environment, Health and Safety Experts were given basic training in the survey audits to ensure a more effective assessment of the locations and demonstrate to the Line Managers the commitment to implement the new *guidelines*.

7 SITE IMPROVEMENT PROGRAMMES

The most important target for the "Responsible Care" programme was to take the comments of the audit teams and transpose them into a site improvement programme.

As head of this new UK Group Environment, Health and Safety function, the author sat down with the head of each location and the Environment, Health and Safety section manager representative and devised an action plan showing what was to be done, by whom and by what target date to meet the new guidelines for the UK group.

8 SUMMARY OF THE ENVIRONMENT, HEALTH AND SAFETY SYSTEM

The new UK Group Environment, Health and Safety System consisted of: *policy, standards, audit* and *improvement programmes*. These Successfully united the many and varied UK operations into a common goal of achieving steadily improving performance - the key to any "Responsible Care Programme" in action. BASF had also embarked on several new initiatives within the auspices of its interpretation of a "Responsible and Caring Company".

The areas most important to all businesses are: *customers, business partners,* and *employees*. It is BASF's belief that customers want products that do not damage the planet, they will prefer goods produced without excessive pollution and will prefer to do business with companies that have a positive image. A process has started of receiving questionnaires from a number of environmentally aware customers who want to know about our managements systems and will want to link together with the best environmental performers spurred on by similar BS5750 quality linkages. Similarly BASF needs to know just how good their business partners' (suppliers and contractors) environmental improvement programmes are and if they have proven credentials. BASF will be striving to secure only the best partners in the 1990's.

Of course, the most important resource is employees, with whom BASF must share their concerns and keep them informed so that good staff will continue to be attracted and retained; this will ensure that they become effective ambassadors for "Responsible Care" programmes.

To promote the debate with local communities and employees, the Chemical Industry has another important element in its "Responsible Care" initiatives: that of the "Speak out and listen" programme. Traditionally, the Chemical Industry has regarded the Public as unlikely to understand the complexity of its business and has seriously neglected its responsibilities in this area. Hopefully it is not too late to intensify this process of caring about the public's concerns and finding ways to demonstrate and communicate, in simple understandable terms, the benefits as well as the risks to them from the Industry. At BASF a number of trained speakers are available to talk on several such topics (Appendix 3).

So far BASF has concentrated its efforts on the local schools where the audiences are always receptive, curious and often searching in their questions. Audiences as diverse as Young Farmers, Rotarians and future physiotherapists at Scarborough Technical college have been addressed! It is a process of gradually chipping away at prejudice, misunderstanding and restoring the debate to one of balance. Not as the media would suggest all risk but also the many benefits often taken for granted.

To conclude the discussion on Responsible Care for the Environment, a description of a recent incident involving a severe storm in the North Sea is relevant. Two tankers of ethyl acrylate were washed overboard from the ferry Nordic Pride, en route from Zeebrugge to Immingham on the morning of Monday, 6 May 1991.

The situation led to the evacuation of three coastal villages, the closing of a five mile stretch of beach and a protracted incident that lasted for three days. It was also an incident that was to bring together not only the Police, Fire and Ambulance services, but encompassed the Coastguard, Royal Air Force, Army Civil Engineering, Marine Surveyors, Shipping Agents both nationally and internationally as well as the District and County Councils. BASF whose material it was that was being transported from its plant in Ludwigshafen West Germany to Immingham on the Humber; also played a major role in providing technical help and assistance with the health care of the communities affected and the local and national media who were quickly on the scene.

During the preceding weekend it was reported that four iso tank containers had been washed overboard from the Swedish registered ship the "Nordic Pride" in a force nine gale in the North Sea. Two that came ashore were carrying ethyl acrylate, a potentially hazardous material used to make acrylic polymers and that this was to become a true test for the "key man" emergency procedure, an important part of the "Responsible Care" programme.....

9 THE INCIDENT

The two tank containers were found by a local resident walking his dog on an isolated stretch of beach at around 0730 hrs on Monday, 6 May 1991. After going back for his camera to take some photographs he informed the Police and Coastguard who were then quickly in attendance.

No identification marks were obvious and there appeared to be a strong smell of what was described as 'garlic' in the air, a strategic retreat was made and the Norfolk Fire Service were contacted at 0824 hrs.

On arrival the Sub Officer, to use a brigade expression, "made pumps two" and requested their chemical incident unit. The crew wearing chemical protection suits and breathing apparatus undertook a search of the containers which were some 200 metres apart to try to identify the contents. They were hampered by the containers laying on their sides and the sea lapping in around the tanks.

Although the labels had been torn off with some of the outer cladding by the severe battering of the tanks at sea, the specially constructed containers appeared to be intact. The contents were subsequently identified by their container numbers and the CHEMSAFE system operated by the Chemical Industries Association triggered the involvement of a member company nearest to the scene of the incident. In this case Dow Chemicals at Kings Lynn, who were incidentally owners of one of the other containers that failed to come ashore on this occasion. Initial information on the hazardous nature of the material was obtained by the Fire Brigade through their chemical incident unit in contact with the national Emergency Centre at Harwell. The next three hours saw the situation escalate.

All personnel were withdrawn and the police and coastguard began to secure the area, prohibiting sightseers and warning local residents to close doors and windows and stay indoors. The wind was northerly, straight off the sea blowing between 10 and 15 knots; this was to cause problems for all involved as the only access was directly into the wind. Further complications arose owing to the terrain in that part of the coast was made up of a steeply sloped shingle beach separated from marshy farm land by a 3 to 4 m high shingle sea defence wall. The only access from the coast road being via a winding track designed for use by the local farmers.

At 1050 hrs the smell in the air was reported to be much stronger and the police decided to evacuate the area as far as Muckleborough, a mile to the east, but this was later extended and an incident room was established in Sheringham Police Station. The nearby villages of Kelling, Weybourne and Salthouse within a two mile radius of the containers were evacuated; this involved 1 000 people. As numerous calls were being received from members of the public said to be suffering from chest, throat and eye irritation, the incident room was relocated to Sheringham High School where additional communication had to be provided.

As BASF had already despatched, on request, a technical support or "Key Man", Mike Groombridge, from the Cheadle office and Martin Penny from the External Affairs department, who happened to live near the scene in Norwich, to the incident area in North Norfolk.

At BASF's base office in Cheadle, the company also established its own incident control room staffed with senior managers and Health and Safety advisors, who kept in constant touch with the "Key Man" in the field with mobile phones.

Back on the beach in North Norfolk, the Emergency Services had decided to try to prevent the containers being carried back out to sea by the tidal waters so an attempt was made to secure them. The operation was to take 1.5 hours and required hawsers to be made fast between the containers and two bulldozers. Civilian engineers were given lessons in wearing protective clothing and breathing apparatus and were accompanied by trained firemen on a one to one basis. The leak at this stage was identified as coming through the safety valve on the roof of the container. Advice provided by BASF's technical expert helped stem the leak to a small drip.

Day 1 of the incident was coming to an end when the BBC 9 o'clock news made the incident item number 2, relegated only by a suspected heart attack to the American President, George Bush.

During the afternoon and evening, Press briefings were organised by the Police assisted by BASF's Press Officer; the media were given all the relevant information to meet their deadlines for reporting the incident. The evacuation of local inhabitants and the treatment at local hospitals of a few people for minor respiratory problems, with no apparent after effects, was sufficient to keep the story going until Day 2.

A comprehensive survey had revealed no significant air pollution and the emergency services faced with the very difficult task of dealing with the tankers from this isolated beach decided to go ahead now without the need for protective equipment unless direct contact with the containers was necessary. They analysed the options as:

1) To refloat the containers out to sea and tow them to a designated port.

2) To decant the contents into road tankers and move the containers on low loaders.

3) To remove the tankers by helicopter bearing in mind the weight of each.

After all parties had deliberated at the debrief session, it was decided that option 2, to decant and move should be adopted. However, it was recognised that low loaders could not get within two miles of the incident and the tankers would get bogged down when filled.

It was then decided that this was now a job for the military and the Army was contacted to see if they could lay a track road and breach the sea defence wall in two places adjacent to each container. The Army were in place by 1845 hrs and rapidly laid an improvised road with their specialist vehicles.

Meanwhile, BASF had been in touch with the parent operation in Germany and had provided medical advice to physicians in Cromer and surrounding hospitals, checked the physical, marine and biological effects of ethyl acrylate with experts in Ludwigshafen and contacted waste disposal operators for assistance.

The speediest and most efficient response came from Leigh Environmental Limited, a reputable waste disposal group, who were quickly on the scene to pump the material into one of their special vehicles for subsequent disposal by incineration. It was later discovered that less than 500 litres out of 24 000 litres had appeared to leak out of the containers, but because of the very low threshold level of smell with this material the "better safe than sorry" arguments had prevailed throughout the incident.

10 IN SUMMARY THEN.....

1) A difficult situation was successfully concluded after almost three days of intense and dedicated effort.

2) The media, including local press and TV networks, unable to get to the site until the afternoon of Day 2 were incessantly requesting information and updates and as a consequence regular interviews had to be set up throughout the incident. The Press Officer from BASF was able to contribute successfully in reporting the simple facts and avoiding the inevitable wild speculation that would otherwise undoubtedly have developed.

3) The technical support provided initially by the "chemsafe" contact in North Norfolk, Dow Chemicals and later by the successful BASF "key man" emergency procedure was excellent even when acts of God at sea produced an unexpected incident ashore.

4) Safety was a prime concern. It could be argued, in hindsight, that the emergency services over-reacted by evacuating over 1,000 people from their homes, but it is always better to scale down an incident once people are in a safe haven. The communities themselves seem little affected after the incident and some would say the renewed community spirit in the area had not been released in this form since the War years.

5) Investigations are continuing by all the parties involved from shipping company, hauliers, insurance companies, emergency services, the chemical industry and government.

The initial indications are that improvements are possible by all the participants and in particular the safe stowage practices of the NorthSea Ferry Operators.

BASF has supported the idea of fundamental changes in the method of securing iso containers on board these ships and will continue to press for improvements in the future to prevent this type of incident occurring again, even in the hazardous conditions that frequently prevail in the North Sea.

Finally, the actions taken during this incident and their successful outcome together with future improvements planned are a further demonstration of today's subject - namely; "Responsible Care of the Environment" which is so important for the future success of the chemical industry and to the survival of the planet for future generations to come.

APPENDIX 1

BASF UNITED KINGDOM ENVIRONMENTAL POLICY

1 STATEMENT

It is our intent to manufacture and distribute products which can be produced, used and disposed of safely and to obtain and provide information on the properties of our products in order to prevent possible harm to people, wildlife and the environment.

BASF Group Companies in the United Kingdom will provide customers and consumers with information and support to enable them to handle and process BASF products in a safe and environmentally acceptable manner.

Sound environmental principles will be practiced in order to comply with laws and governmental regulations. Furthermore Group Companies will pursue responsible improvement programmes and impliment additional measures necessary to protect the environment even though not mandated by regulations or legislation.

An open exchange of information will be undertaken with governmental agencies, organisations and the public on environmental protection and safety.

2 ORGANISATION RESPONSIBILITIES

The Group Environment, Health and Safety Manager is responsible for maintaining environmental guidelines for BASF Group Companies in the United Kingdom.

Company Managing Directors and Location Heads are responsible for developing and maintaining environmental procedures within the guidelines appropriate to each location and providing information to their employees, customers and the public.

APPENDIX 2

ENVIRONMENT, HEALTH AND SAFETY GUIDELINES (September 1990)

<u>Section A</u>

1. General Introduction
2. Worldwide BASF Group Environmental Protection and Safety Guidelines.
3. Environment, Health and Safety Policy Statements
4. Environment, Health and Safety Procedures Manuals

<u>Section B:</u> Health and Safety Guidelines

1. Legislation and Standards
2. Safety Committees
3. Induction Training
4. Permits and Clearance Certificates
5. Fire Detection and Protection Systems and Fire Emergencies
6. First Aid
7. Reporting and Investigation of Incidents/Injuries
8. Personal Protective Equipment
9. Process Safety and Control
10. Emergency Planning
11. Transport of Chemicals
12. Warehousing and Storage
13. Product Safety Data
14. Environment, Health and Safety Audits
15. Electrical Safety at Work
16. Noise and Hearing Conservation
17. Control of Substances Hazardous to Health (COSHH)
18. Work Place Monitoring
19. Manual Handling
20. Display Screen Equipment (VDUs)
21. Control of Contractors

APPENDIX 3

THE CHEMICAL INDUSTRIES ASSOCIATION "SPEAK OUT" PROGRAMME: PUTTING CHEMICALS INTO PERSPECTIVE

The good the chemical industry does for the individual and for the country too often goes by default. This is partly the industry's own fault as it has not always said enough about what it does, how carefully it operates, its products' beneficial contribution to modern life, and its positive contribution to the economy in terms of jobs, investment and the balance of payments.

The new "Speak Out Programme", sponsored by the Chemical Industries Association, sets out to put this right. It involves a team of speakers, all experts in different aspects of the industry, one of whom will go to talk to any club, society or educational establishment in the United Kingdom, free of charge. The presentations, which use specially prepared 35mm slides, are based on the list of subjects shown on the opposite page, enlivened by the speaker's own experience. Subjects can be combined , and designed to meet any special requirements needed by the group. The choice is the recipients, as is the time of day, and day of the week.

Response so far is most encouraging. A face-to-face meeting with a member of the industry really does put chemicals into perspective. Students have found the talks directly useful in providing an insight into the practical application of science which is in line with the latest curricular requirements. Adult audiences have shown great interest in the up-to-date information on current issues, and in both the history and the future of the industry. Generally they report that their opinion of the industry has been enhanced.

"Speak Out" can promise an entertaining, thought-provoking and stimulating discussion, plus solid information. There is no ducking and weaving. The speaker is there not to defend the industry come what may, but to promote a better understanding of its role. Questions will be answered, and points of view noted and reported back. Audience participation is highly valued.

SUBJECTS TO CHOOSE FROM

Presentations can be based on a single subject, or a blend of two or more.

In The Home

Chemicals are everywhere around the home, making life easier, more colourful and safer. This illustrates their varied roles in d-i-y, gardening and other leisure activities, in the basic structure and services of the house, in furnishing, cars and communications.

Healthcare

Modern healthcare is largely dependent on chemicals; without them life expectancy would be significantly lower. "Healthcare" deals with sanitation, cleaning and disinfection, food production and storage, hospital care and medicines and modern clothing.

Safety and the Environment

The chemical industry has a significant impact on the environment. This presentation shows how it minimises hazard through sound plant design, training and operations management. It also deals with waste disposal and the industry's handling of road transport.

Economic Contribution

The chemical industry is the UK's fourth largest industry and vital to many others: engineering, farming, food and electronics, to name but a few. Slides illustrate its positive contribution to the balance of trade and to employment and investment. Its profitability enables it and its staff to make substantial payments for the benefit of the community at large.

The Future

In a research-based industry, chemicals are likely to make an ever growing contribution to our future. Innovation will be stimulated by human and industrial needs for cost reduction, better performance and lower environmental impact and by basic scientific advances involving new materials and processes. The presentation discusses how the industry goes about its research work and describes some of the most promising areas of activity.

Chemicals Through The Ages

Man has used chemicals for centuries, deriving them in early times from natural materials like wood ash, plants and roots. Today scientific understanding has so improved that products can be created virtually to order. This talk traces the evolution of primitive cures and colours to modern pharmaceuticals and dyes and of processes from the single crude furnace to today's computer-controlled operations.

Subject Index